Thinking Evolutionarily

Evolution Education Across the Life Sciences

Summary of a Convocation

Steve Olson, *Rapporteur*
Jay B. Labov, *Editor*

Planning Committee on Thinking Evolutionarily:
Making Biology Education Make Sense

Board on Life Sciences, Division on Earth and Life Studies

NATIONAL RESEARCH COUNCIL
OF THE NATIONAL ACADEMIES
NATIONAL ACADEMY OF SCIENCES

THE NATIONAL ACADEMIES PRESS
Washington, D.C.
www.nap.edu

THE NATIONAL ACADEMIES PRESS 500 Fifth Street, NW Washington, DC 20001

NOTICE: The project that is the subject of this report was approved by the Governing Board of the National Research Council, whose members are drawn from the councils of the National Academy of Sciences, the National Academy of Engineering, and the Institute of Medicine. The members of the committee responsible for the report were chosen for their special competences and with regard for appropriate balance.

This study was supported by the National Academy of Sciences and grants from the Burroughs-Wellcome Foundation, Christian A. Johnson Endeavor Foundation, a Research Coordination Network/Undergraduate Biology Education Grant from the National Science Foundation to Oklahoma University, and in-kind support from the Carnegie Institution for Science. Any opinions, findings, conclusions, or recommendations expressed in this publication are those of the author(s) and do not necessarily reflect the views of the organizations or agencies that provided support for the project.

International Standard Book Number-13: 978-0-309-25689-6
International Standard Book Number-10: 0-309-25689-5

Additional copies of this report are available from the National Academies Press, 500 Fifth Street, NW, Keck 360, Washington, DC 20001; (800) 624-6242 or (202) 334-3313; Internet, http://www.nap.edu.

Printed in the United States of America

Suggested citation: National Research Council and National Academy of Sciences (2012). *Thinking Evolutionarily: Evolution Education Across the Life Sciences. Summary of a Convocation.* Steve Olson, Rapporteur. Planning Committee on Thinking Evolutionarily: Making Biology Education Make Sense. Board on Life Sciences, Division on Earth and Life Studies, National Research Council, and National Academy of Sciences. Washington, DC: The National Academies Press.

PLANNING COMMITTEE ON THINKING EVOLUTIONARILY: MAKING BIOLOGY EDUCATION MAKE SENSE

CYNTHIA M. BEALL* (*Chair*), Department of Anthropology, Case Western Reserve University
PAUL BEARDSLEY, Department of Biological Sciences, California Polytechnic University, Pomona
IDA CHOW, Society for Developmental Biology
JAMES P. COLLINS, School of Life Sciences, Arizona State University
IRENE ECKSTRAND, National Institute of General Medical Sciences, National Institutes of Health
KRISTIN JENKINS,** Education and Outreach, National Evolutionary Synthesis Center
NANCY A. MORAN,* Department of Biology, Yale University

GORDON E. UNO,† Department of Botany and Microbiology, Oklahoma University

JAY B. LABOV, Senior Advisor for Education and Communication and Study Director
CYNTHIA A. WEI, Christine Mirzayan Policy Fellow, National Academy of Sciences
ORIN E. LUKE, Senior Program Assistant

---

*Member, National Academy of Sciences.
**Current Affiliation: BioQuest.
† Special Consultant to the Organizing Committee.

BOARD ON LIFE SCIENCES

JAY LABOV, Senior Advisor for Education and Communication
KEEGAN SAWYER, Program Officer
MARILEE SHELTON-DAVENPORT, Senior Program Officer
AYESHA AHMED, Senior Program Assistant
CARL-GUSTAV ANDERSON, Program Associate
ORIN LUKE, Senior Program Assistant

Acknowledgments

This workshop summary is based on discussions at a convocation that was organized by a committee under the aegis of the Board on Life Sciences of the National Research Council (NRC) and the National Academy of Sciences on October 25-26, 2011. We thank our colleagues who served on the planning committee, each of whom brought critical expertise and perspectives to the planning of the convocation. The planning committee members identified speakers and panelists, helped organize and finalize the agenda, and facilitated discussions during the two breakout sessions. Several committee members also served as panelists during the convocation (see Appendix A). Although the committee was neither tasked with nor contributed to the writing of this summary, this publication clearly reflects its diligent efforts along with the excellent presentations by experts, and the insightful comments of the many participants during the convocation.

This convocation would not have been possible without the generous support of the Burroughs Wellcome Fund, the Christian A. Johnson Endeavor Foundation, the National Academy of Sciences, and the National Science Foundation through a Research Coordination Network/ Undergraduate Biology Education grant to Oklahoma University (Gordon Uno, Principal Investigator). We thank all of them sincerely. We also thank Dr. Toby Horn, Carnegie Institution for Science, for her role in procuring the facilities of the Carnegie Institution for the convocation and in assisting with logistical planning for the event.

This summary has been reviewed in draft form by individuals chosen for their diverse perspectives and technical expertise, in accordance with procedures approved by the NRC's Report Review Committee. The purpose of this independent review is to provide candid and critical comments that will assist the institution in making its published summary as sound as possible and to ensure that the summary meets institutional standards for objectivity, evidence, and responsiveness to the study charge. The reviewers' comments and draft manuscript remain confidential to protect the integrity of the process. We thank the following individuals for their review of this summary:

Clarissa Dirks, The Evergreen State College
Adam Fagen, Genetics Society of America
David Jablonski, University of Chicago
Kenneth R. Miller, Brown University
Elvis Nuñez, University of Florida
Paul Strode, Fairview High School, Boulder, CO
David Wise, University of Illinois, Chicago

Although the reviewers listed above provided many constructive comments and suggestions, they were not asked to endorse the content of the report, nor did they see the final draft of the report before its release. The review of this report was overseen by Dr. Diane Ebert-May, Michigan State University. Appointed by the NRC, she was responsible for making certain that an independent examination of this report was carried out in accordance with institutional procedures and that all review comments were carefully considered. Responsibility for the final content of this report rests entirely with the author and the institution.

We are grateful for the leadership and support provided by Kenneth R. Fulton, executive director of the National Academy of Sciences, and Frances Sharples, director of the NRC's Board on Life Sciences. We thank Orin Luke, senior program assistant, for his valuable contributions to planning and implementing the logistics for all aspects of the convocation. We also thank Rebecca Fischler, communications officer in the NRC's Division on Earth and Life Studies, for her critical expert advice and assistance with developing and maintaining the convocation's website (*http://nas-sites.org/thinkingevolutionarily/*) and electronic procedures.

We acknowledge the important contributions of the National Evolutionary Synthesis Center for organizing and supporting the working committee that envisioned this convocation and the role of the NRC and National Academy of Sciences as the convening bodies for the event (for additional information about this project, see *http://nas-sites.org/thinkingevolutionarily/convocation-description/*).

Finally, we thank all of the participants for taking the time and, for many, the expense to attend this convocation. We are also deeply grateful to the following disciplinary and professional societies for sending representatives to the convocation: American Association for the Advancement of Science, American Institute for Biological Sciences, American Society for Microbiology, American Society of Human Genetics, American Society of Plant Biologists, American Society of Primatologists, Animal Behavior Society, Association of American Medical Colleges, Biophysical Society, Ecological Society of America, Entomological Society of America, Federation of American Societies for Experimental Biology, Human Anatomy and Physiology Society, National Association of Biology Teachers, National Science Teachers Association, Phycological Society of America, Society for Developmental Biology, Society for Freshwater Science, Society for Integrative and Comparative Biology, and the Society for the Study of Evolution.

Cynthia M. Beall, Ph.D.
Chair, Organizing Committee

Jay B. Labov, Ph.D.
Study Director and Editor

Contents

APPENDIXES

1

Introduction and Overview[1]

Evolution is the central unifying theme of biology. Yet today, more than a century and a half after Charles Darwin proposed the idea of evolution through natural selection, the topic is often relegated to a handful of chapters in textbooks and a few class sessions in introductory biology courses. In many introductory biology courses (both undergraduate and high school), and even in some upper-level courses, evolution is not covered at all.

In recent years, a movement has been gaining momentum that is aimed at radically changing this situation. An increasing number of research scientists, educators, and education researchers are pointing to the many benefits of teaching evolution throughout the biology curriculum. Understanding evolutionary processes is essential to achieving a full understanding of the variety, relationships, and functioning of living things. An appreciation of evolutionary principles can enhance and enliven study of virtually all other areas of biology, such as embryological development, the spatial distribution of organisms, anatomy and physiology, behavior, interactions among organisms, processes of disease, the biological history of all species including humans, and a greater appreciation for biodiversity and the natural environment. Furthermore, teach-

[1] This report has been prepared by the workshop rapporteur as a factual summary of what occurred at the workshop. The planning committee's role was limited to planning and convening the workshop. The views contained in the report are those of individual workshop participants and do not necessarily represent the views of all workshop participants, the planning committee, or the National Research Council.

ing evolution across the curriculum can help counter the confusion and contention that still hinder the teaching of evolution in many classrooms, especially at the K-12 level, in the United States.

On October 25-26, 2011, the Board on Life Sciences of the National Research Council and the National Academy of Sciences held a national convocation in Washington, DC, to explore the many issues associated with teaching evolution across the curriculum. Titled "Thinking Evolutionarily: Evolution Education Across the Life Sciences," the convocation brought together people from many sectors, including K-12 education, higher education, museums, publishers, government, philanthropy, international educators, and non-profit organizations, who rarely communicate but need to work collaboratively if evolution is to assume a more prominent role in biology education. The goals of the convocation were to articulate issues, showcase resources that are currently available or under development, and begin to develop a strategic plan for engaging all of the sectors represented at the convocation in future work. It focused specifically on infusing evolutionary science into introductory college courses and into biology courses at the high school level, although participants also discussed learning in earlier grades and life-long learning. In addition, the convocation covered the broader issues associated with learning about the nature, processes, and limits of science, because understanding evolutionary science requires a more general appreciation of how science works.

This summary provides a narrative, rather than a chronological, overview of the presentations and rich discussions that occurred during the convocation. It is organized around the major themes that recurred throughout the event, including the structure and content of curricula, the processes of teaching and learning about evolution, the tensions that can arise in the classroom, and the target audiences for evolution education.

For a much more complete list of resources, see the annotated bibliography that is found in *Science, Evolution, and Creationism* (National Academy of Sciences and Institute of Medicine, 2008) and the resources found throughout the National Academy of Sciences' *Evolution Resources* webpage (*http://nationalacademies.org/evolution*). In addition, resources that were suggested prior to and following the convocation by planning committee members and participants can be found at *http://nas-sites.org/thinkingevolutionarily/additional-resources/*.

THE SETTING AND SPIRIT OF THE CONVOCATION

The convocation was held at the Carnegie Institution for Science in Washington, DC, which has supported major science initiatives throughout the 20th and 21st centuries. In her welcoming remarks at the convocation, Maxine Singer, a member of the National Academy of Sciences and Institute of Medicine, and President Emerita of the institution, recalled her service on the committee that wrote the first edition of the report *Science and Creationism* (National Academy of Sciences, 1984). The committee's meetings were enlivened by the exchanges of two accomplished physical scientists, she said. "One, an adamant, feisty, and cerebral non-believer, would have preferred us to offer bold language that set religion aside as a way to view the world. The other, a calm and at least as cerebral religious believer who was also firmly convinced by the evidence for biological evolution, urged us toward an understanding and tolerance of religion."

The committee listened carefully to this discussion, Singer said, and what it learned is captured in the eloquent conclusion to the 1984 report: "Scientists, like many others, are touched with awe at the order and complexity of nature. Religion provides one way for human beings to be comfortable with these marvels. However, the goal of science is to seek naturalistic explanations for phenomena within the framework of natural laws and principles and the operational rule of testability."

This is the spirit in which the convocation was held. "My hope," said Singer, "is that we all respect the religious beliefs of one another, of students and their families. I think you can find ways to teach evolution that are scientifically rigorous but avoid contentious challenges to individuals."

PERSPECTIVE OF A FUNDER

The convocation was funded by the National Academy of Sciences, the Burroughs Wellcome Fund, the Christian A. Johnson Endeavor Foundation, the Carnegie Institution for Science, and the National Science Foundation through a Research Coordination Network/Undergraduate Biology Education grant to the University of Oklahoma. A representative of one of the funders, Susan Kassouf, a program officer at the Johnson Endeavor Foundation, spoke in the opening session about some of the larger issues addressed during the convocation. She said that the mission of the Johnson Endeavor Foundation is to help people, especially young people, flourish. It has pursued this mission by helping to provide students with a liberal arts education that offers the best thinking of humanity. For this reason, among others, the foundation has become interested in understanding why so many Americans doubt evolutionary science when such doubt can have grave consequences not only for the individual but also for the larger society.

"Getting one's head, heart, and soul around the scientific theory of evolution and its implications is daunting," said Kassouf. "While our awe and wonder about the world may deepen in light of evolutionary theory—indeed, evolution does seem miraculous—our minds may also boggle and buckle when coming to terms with a certain fundamental randomness and unpredictability, a lack of a grand design, a perception that the theory portends a loss of meaning and purpose in our lives. For all of these reasons and others, we applaud your efforts to make the scientific theory of evolution an integral part of young people's introduction to biology and help them become comfortable with this fundamental, perhaps unsettling, idea."

The theory of evolution can be seen to underlie our entire understanding of life, said Kassouf. Efforts such as the ones being discussed at the convocation are "a wise way to help us all begin to accept the soundness of evolutionary theory not just in our heads but in our hearts and minds."

OVERVIEW OF THE CONVOCATION[2]

In his opening presentation, Gordon Uno, David Ross Boyd Professor at the University of Oklahoma, as well as a member of a group under the National Evolutionary Synthesis Center (NESCent) that first conceived of this convocation and a special consultant to the convocation's organizing committee, laid out many of the central issues addressed at the event.

Teaching evolution across the curriculum makes sense both biologically and pedagogically, he said. (Chapter 2 describes some of the many curricular and instructional changes needed to teach evolution across the curriculum.) Many major science education reform movements have observed that students learn better when information is organized around major unifying concepts such as evolution (see Box 1-1). In biology, no concept is more unifying than evolution. The biologist Theodosius Dobzhansky wrote an article with the famous title, "Nothing in Biology Makes Sense Except in the Light of Evolution" (Dobzhansky, 1973). Uno offered a corollary: Everything in biology makes more sense in the light of evolution. "If we really want to help our students understand biology, shouldn't we be teaching more evolution?"

Instructors and students should clearly understand the learning objectives for a course, Uno observed. Instructors then should ask what activities, lessons, and other experiences will help students reach those objectives. In this way, teachers have a constant reminder to be intentional in their instruction.

For the biology course he teaches, Uno's reminder is: "Evolution—say it every day." It is a challenge to incorporate something about evolution in every class taught in every course. But when Uno talks about cells, he

---

[2] Additional resources, including video archives and PowerPoint presentations of speakers and panelists, interviews with selected participants, and a list of useful references and websites are available at *http://nas-sites.org/thinkingevolutionarily/*.

BOX 1-1
Prominent Statements on Evolution Education

From the *National Science Education Standards* (National Research Council, 1996): As a result of activities in grades K–12, all students should develop understanding and abilities aligned with the following concepts and processes:

- Systems, order, and organization
- Evidence, models, and explanation
- Constancy, change, and measurement
- Evolution and equilibrium
- Form and function

From *Scientific Foundations for Future Physicians* (AAMC-HHMI, 2009), Competency E8: Demonstrate an understanding of how the organizing principle of evolution by natural selection explains the diversity of life on earth.

From *A Framework for K-12 Science Education* (National Research Council, 2011): Biological evolution explains both the unity and diversity of species and provides a unifying principle for the history and diversity of life on Earth.

From *Vision and Change in Undergraduate Biology Education: A Call To Action* (Brewer and Smith, 2011): The diversity of life evolved over time by processes of mutation, selection, and genetic change.

From *AP Biology Curriculum Framework: 2012-2013* (College Board, 2011): The process of evolution drives the diversity and unity of life.

From *Science, Evolution, and Creationism* (National Academy of Sciences and Institute of Medicine, 2008): Biological evolution is the central organizing principle of modern biology.

looks for opportunities to talk about endosymbiosis. When he teaches about photosynthesis, he provides evolutionary explanations for why plants do not absorb the green part of the visible spectrum and thus reflect green wavelengths of light. When he talks about cells, he also describes the same sorts of molecules in different organisms and the relevant evolutionary history. "Every single day I try to bring into the classroom something about evolution."

As described in Chapter 3, education researchers still have much to learn about how students learn evolution and about the effects of an evolutionary understanding on other aspects of biology education. But Uno listed several questions that he asks students to gauge whether they are thinking evolutionarily:

- How did that evolve?
- Is that the same in all organisms?
- What is the significance of that structure?
- How can that be explained?
- How does this process or phenomenon compare to that one?
- Is this biologically related to that?
- What does that information tell us about the evolution of X?
- How does one develop curricular material that gets to everyone?

Many questions surround instruction and the development of supporting curricular materials for evolution education. Are there ways to teach all students critical concepts in evolutionary science such as artificial and natural selection, emerging diseases, developmental biology, key transitions in the history of life, biodiversity, or evolutionary medicine? Who should develop the materials needed to teach these concepts, and how can biologists be convinced to contribute to their development? How can people be made aware of these materials and be convinced to use them? And how can the effects of materials and instructional approaches be measured? All of these questions are potential subjects of research.

Teaching evolution across the curriculum also can thwart the constant assault on the teaching of evolution (Chapter 4). "I'm from Oklahoma. We are the buckle on the Bible belt, and I deal with a lot of students on a regular basis in my introductory courses who show resistance to teaching and accepting evolution." In high schools in Oklahoma and throughout the nation, students are often absent on the days when evolution is taught, Uno stated. Even in colleges, when evolution is listed on the schedule, students miss those days. "If you teach evolution every single day, then there is no avoiding evolution," said Uno.

Uno encouraged the convocation participants to think outside the box about target populations, which is the subject of Chapter 5. High school students and teachers are major audiences of course. But can ways be found to reach farmers, parents, and politicians? Farmers understand selection, because they understand the evolution of pesticide resistance as well as how much their crops and livestock can be changed over time through selective breeding. "Is there a way that we can reach that population by customizing our information or our message?" asked Uno. Parents could be receptive to a message about emergent diseases. Other important audiences include faculty and students at two-year colleges, textbook authors and publishers, and media people. "We need to think about customizing our message and our strategies for individuals at these different kinds of institutions." To reach a broad spectrum of audiences, both top-down and bottom-up public relations campaigns will be needed.

Uno was a member of the Evolution Across the Curriculum (EVAC)

Working Group sponsored by NESCent that initially proposed the convocation.[3] (Chapter 6 describes the progress that has been made to date in implementing the idea and the resources available to make continued progress.) The working group did not set out to produce a curriculum per se. Rather, it focused on compiling and developing materials for instructors at the high school and undergraduate levels. It also examined how to get instructors to contribute and use evolutionary examples in their teaching and how to get people to think evolutionarily.

In addition, Uno has been part of a group that has been revising Advanced Placement (AP) Biology based on recommendations from a National Research Council report (2002). The updated course, which is scheduled to be implemented during the 2012-2013 school year, is framed in terms of four big ideas or unifying themes, and the first one concerns evolution: "The process of evolution drives the diversity and unity of life." The AP Biology exam also is being revised to have a much greater emphasis on evolution. Uno found that in a released version of the 2008 exam, 12 percent of the questions had something to do with evolution. In the exam being developed for the restructured course, at least 35 percent of the questions will involve evolution (based on a recently released practice examination). "That was not a target. We didn't say, 'We need to have X number of questions that have something to do with evolution.' This is a natural consequence of framing a course that uses evolution as one of the themes."

Finally, Uno described some of the steps needed to make accelerated progress in teaching evolution across the curriculum, which is the subject of the final chapter (Chapter 7) in this summary report. Many of these steps involve more than curricula and teaching materials; they depend on the attitudes of and relationships among scientists, teachers, students, and the public. A public relations campaign is essential, he said. When students and parents say, "Teach the controversy," "Give equal time to creationism," or "Evolution is not based on sound science," instructors of biology need ready counter-statements. Uno suggested that a powerful statement for the general public is, "That's just another example of evolution in action."

By bringing together like-minded people from different backgrounds, the convocation was designed to create enduring collaborations, Uno pointed out in concluding his remarks. The idea was to find out what

[3] Two other members of the EVAC working group, Paul Beardsley and Kristin Jenkins, served as members of the organizing committee. A third member of the EVAC group, Jay Labov, was the staff officer who directed the project. Additional information about this project is available on the convocation (*http://nas-sites.org/thinkingevolutionarily*) and NESCent websites (*http://www.nescent.org*).

works and what does not work to help students learn biology with an evolutionary perspective, and then to institute what works. Who are the key players in promoting the teaching of evolution? How can those who need assistance to teach more evolution in their courses get that assistance, and how can that assistance be delivered? If curricular materials are the answer, who is going to develop them, and how will people learn about them? These are among the many questions the convocation was meant to examine, Uno said.

REFERENCES

AAMC-HHMI (Association of American Medical Colleges-Howard Hughes Medical Institute). 2009. *Scientific Foundations for Future Physicians.* Washington, DC: Association of American Medical Colleges.

Brewer, C., and Smith, D, Eds. 2011. *Vision and Change in Undergraduate Biology Education: A Call to Action.* Washington, DC: American Association for the Advancement of Science.

College Board. 2011. *AP Biology Curriculum Framework: 2012-2013.* Princeton, NJ: College Board.

Dobzhansky, T. 1973. Nothing in biology makes sense except in the light of evolution. *American Biology Teacher* 35:125-129.

National Academy of Sciences. 1984. *Science and Evolution: A View from the National Academy of Sciences.* Washington, DC: National Academy Press.

National Academy of Sciences and Institute of Medicine. 2008. *Science, Evolution, and Creationism.* Washington, DC: The National Academies Press.

National Research Council. 1996. *National Science Education Standards.* Washington, DC: National Academy Press.

National Research Council. 2002. *Learning and Understanding: Improving Advanced Study of Mathematics and Science in U.S. High Schools.* J. P. Gollub, M. Bertenthal, J. Labov, P. C. Curtis, Eds. Washington, DC: National Academy Press.

National Research Council. 2011. *A Framework for K-12 Science Education: Practices, Crosscutting Concepts, and Core Ideas.* Washington, DC: The National Academies Press.

2

Changing Curricula and Instruction

Two speakers at the convocation—Robert Pennock, professor at Michigan State University, and Bruce Alberts, a member and former President of the National Academy of Sciences, professor emeritus at the University of California, San Francisco, and editor-in-chief of the journal *Science*—discussed the broad issues involved in teaching evolution across the curriculum. Teachers' acceptance and understanding of evolution can have major impacts on its dissemination into the classroom. In addition, educators often encounter resistance in teaching evolution, and both speakers discussed ways of overcoming this resistance. (Conflicts in the teaching of evolution also are discussed in chapter 4.) Many aspects of instruction and curricula will need to change to make evolution a continual presence in biology education, Pennock and Alberts emphasized, yet these changes could strengthen both biology education and students' grasp of how evolution works and why it is important.

CHALLENGE AND RESPONSE

In 1996, a school superintendent in Kentucky ordered two pages of a textbook glued together because they provided a scientific explanation for the creation of the universe while not also presenting the Bible's explanation. Teaching evolution throughout the curriculum would make it impossible to avoid the subject, said Pennock. The challenge for the convocation, said Pennock, is: "How can we make sure that you couldn't do this unless you had to glue the whole textbook together?"

An Evolving Controversy

Most of the critics of evolution no longer directly challenge the idea of teaching scientific concepts in science classrooms, Pennock noted. Instead, they proclaim that teachers should "teach the controversy." For example, a bill introduced in Michigan a few years ago requires teachers to "(A) use the scientific method to critically evaluate scientific theories including, but not limited to, the theories of global warming and evolution," and "(B) use relevant scientific data to assess the validity of those theories and to formulate arguments for and against those theories."[1] As Michigan Representative John Moolenaar said, this language leaves it up to local school boards whether to require the teaching of intelligent design (ID)—the idea that living things are too complex not to have been created by a divine or supernatural intelligence.

Pennock noted that this approach was soundly repudiated in the federal court case *Kitzmiller et al. v. Dover Area School District*. As the judge in that case wrote, "ID's backers have sought to avoid the scientific scrutiny which we have now determined that it cannot withstand by advocating that the controversy, but not ID itself, should be taught in science class. This tactic is at best disingenuous, and at worst a canard." Yet critics of evolution continue to try to insert religious ideas into science classes using this approach. When intelligent design creationists proposed to the Texas Board of Education that students be required to analyze and evaluate "strengths and weaknesses" in evolutionary theory, the board voted against the proposal, after which creationists proposed that students study "evidence supportive and not supportive of a theory." The board again voted against the proposal, but when creationists next proposed that students study "the sufficiency or insufficiency of common ancestry to explain the sudden appearance, stasis, and sequential nature of groups in the fossil record"—which are all buzzwords for intelligent design creationism—the board accepted the proposal. "It's never quite over," said Pennock. "You have to pay attention to the way that words are used, and language makes a difference."

Using Language Carefully

Especially in teaching evolution, teachers need to be very precise in the language they use, said Pennock, because students and the public are very attuned to the nuances of particular terms. "The way in which we frame these issues can make a difference in terms of whether they're going to be accepted."

[1] Almost identically worded bills have been proposed during the past several years in the legislatures of several other states.

Particularly problematic is the use of the word *theory*. For example, the disclaimer that the Cobb County, Georgia, Board of Education approved to insert into the textbook *Biology* by Kenneth Miller and Joseph Levine (2004) and other texts that include discussion of evolution read, "This textbook contains material on evolution. Evolution is a theory, not a fact, regarding the evolution of living things."[2] But the colloquial meaning of *theory* is very different from the scientific meaning. The general public interprets the word *theory* as a guess or supposition—"you have your theory and I have mine." In science, a theory is a broad overarching explanation that accounts for a wide variety of empirical observations. To avoid this confusion, Pennock uses the term *evolutionary science* rather than evolutionary theory. "This is a way of avoiding a word that we know is going to trip people up," he said.

Many other terms commonly used in evolutionary science have ambiguous or multiple meanings, including *adapt*, *selection*, and even *evolution* itself. These terms need to be defined and used carefully to avoid confusing scientific and colloquial meanings.

Getting Learners Hooked

In countering attacks on evolution, the scientific community tends to be reactive, said Pennock. A legislative proposal needs to be defeated. The statements of a creationist politician need to be countered. From this perspective, the discussion becomes a debate, with each side presenting its best arguments.

The scientific community needs to think about becoming more proactive, said Pennock. In this way, people could be reached before the discussion becomes a debate.

This approach is complicated by the large percentage—approximately 40 percent—of people in the United States who believe that evolution is false (Figure 2-1). Even 32 percent of students with a college education answered "no" to the question, "Do you think that the modern theory of evolution has a valid scientific foundation." In fact, among high school biology teachers, 40 percent think that "there are sufficient problems with the theory of evolution to cast doubts on its validity" (Berkman and Plutzer, 2011).

The best opportunity, suggested Pennock, lies in reaching the 20 percent of Americans who are unsure about the accuracy of evolution. "That has to be a primary target, not initially to reach those who are opposed

[2] The full text of this sticker reads: "This textbook contains material on evolution. Evolution is a theory, not a fact, regarding the origin of living things. This material should be approached with an open mind, studied carefully, and critically considered."

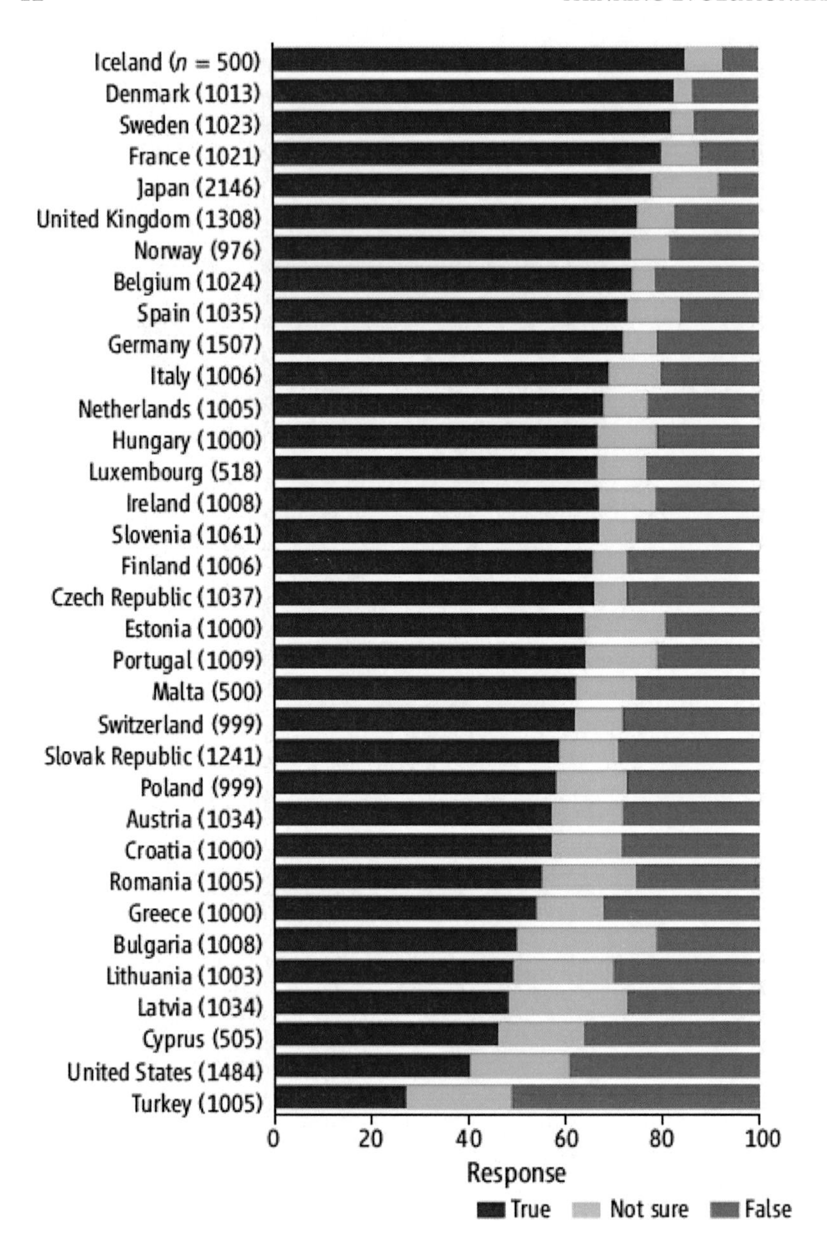

Public acceptance of evolution in 34 countries, 2005.

FIGURE 2-1 A larger percentage of people reject evolution as false in the United States than in almost all other developed countries.

ideologically, but to reach those in the middle who just don't know." It also is particularly important to reach teachers and future teachers, because they are the ones who will teach their students the subject.

Scientists often think that the best way to convince the undecided is to marshal the data. But at the frontiers of knowledge, scientific discussion most often takes the form of evidence-based persuasion, Pennock said. "How do we get students to think about evidence-based reasoning? Clearly, this is where we want to get them."

To think in these terms, students and members of the public first need to be hooked. The way to do that, said Pennock, is to start not with data but with something that gets them interested. Evolutionary science has many topics featuring practical applications, such as evolutionary medicine, pest management, forensic tools, evolutionary computation, and evolutionary engineering design. Most scientists themselves became interested in science because of a hook that got them emotionally engaged, observed Pennock. Only after they were interested in science did they learn about hypotheses, data, and predictions. "The initial thing is how you reach their hearts. Their minds then come next. Data isn't the first thing. We think of data first, but actually data is last. The first thing is how you hook them."

The same observation applies in interacting with the media. Scientists want to dwell on the data, whereas journalists are most interested in why a topic matters. "When they write the story, that's what they'll write first. The lead of the story is the upshot. What's in it for us? Then, once they've hooked you, they can present the data."

As a specific example of how to make evolution relevant, Pennock mentioned evolutionary medicine. Medical students are interested in why people get sick, and if those reasons have evolutionary roots, these students can become interested in evolution. Along the same lines, the education committee of the Society for the Study of Evolution has been holding an annual symposium on applied evolution for the past 10 years. Although the society was initially skeptical, the symposium has become one of the most popular it offers.

BEACON

Pennock described in greater detail a particular way to get students interested in evolution. One complication with teaching evolution is that it has been difficult to do evolutionary experiments in real time, but Pennock and his colleagues have developed ways of doing just that. The Bio/computational Evolution in Action Consortium (BEACON) is a new science and technology center at Michigan State University that explores

BOX 2-1
Thinking Evolutionarily

The first of two sets of breakout groups during the convocation addressed the general question: "What constitutes evolutionary thinking?" Individual breakout groups were asked to address the following issues more specifically:

Group 1: What approaches are needed to educate faculty and departments about the value of evolutionary thinking in their own courses and programs?

Group 2: What additional evidence is needed to convince biologists of the value of evolutionary thinking? How can that evidence best be gathered through an organized program of research? Who should undertake and sponsor such research?

Group 3: How can evolutionary thinking become more firmly connected with other emerging efforts to improve life sciences education? In what ways should these efforts be influenced by different target audiences?

During the plenary reporting session that followed the breakout group meetings, multiple breakout group participants made several main points.

First, evolutionary science is driven by evidence. As a result, there can be uncertainty about specific questions. Evolutionary understanding continues to progress as more questions are answered and as understanding is refined. If students understand that the science continues to advance at the forefront of knowledge, then they can take a big step toward understanding not only how evolutionary biology but also how science in general works.

evolutionary processes in both biological and computational systems.[3] BEACON studies evolution in real time using real organisms in laboratories and field sites and "digital organisms" that evolve in computers. On the biological side, for example, Pennock pointed to Michigan State University's Richard Lenski, who has been conducting a long-term evolution experiment with *E. coli* for more than 20 years. By following evolution in parallel lines of *E. coli* for more than 50,000 generations, he and his colleagues have observed evolutionary adaptations in all of the lines.

On the computational side, random variation, selection, and evolution all can be modeled in computer systems. Using a system called Avida, for example, students can explore basic evolutionary mechanisms and test hypotheses (Box 2-1). In one set of experiments, students can use a virtual Petri dish to observe a model for the evolution of colonies of

[3] Additional information is available at *http://beacon-center.org*.

Evolutionary biology is also probabilistic. The genes that are passed from one generation to the next cannot be predicted with certainty, which means that the future evolutionary pathways cannot be predicted with certainty. Students increasingly need to think about evolution in probabilistic terms as their level of understanding advances.

Evolutionary biology is more dependent on variation than are other sciences. This variation can be both observed and appreciated even by very young students. Even second graders, for example, can observe the differences and similarities among siblings.

Finally, evolutionary biology has a historical dimension. Life on earth evolved over billions of years, which means that unique things have happened in the course of that history. Every hydrogen atom behaves like every other hydrogen atom. But every gene, individual organism, and species is the result of a historical process and cannot be fully understood without understanding that history. Particular lineages have evolved, and these events can probably happen only once because of the unique combination of genes, environments, and chance that gave rise to that lineage. The history of life is contingent, so that if it were rerun, the same things would not likely happen. Nevertheless, there is a tree of life that can never be completely described but can be continually explored.

Students bring many misconceptions to their study of evolution. One is that evolution is a progressive process in which humans are the pinnacle of a long chain of advancement. Another is that everything in nature is optimized because it has evolved to fit perfectly with the environment. A much more accurate concept is that evolution involves tradeoffs between costs and benefits. A big brain has advantages for humans but makes birth more difficult than it is for other mammals. Standing upright has advantages but creates a greater likelihood of back pain.

virtual organisms. As sub-colonies in this virtual system "evolve" new characteristics through the appearance of random mutations, they can take over their predecessors in the colony. Students can vary the mutation rate or the resources available to the virtual organisms. "We finally have a new lever to let students observe [models for mechanisms of evolution] for themselves and do so through inquiry-based lessons."

Intelligent design creationists have been alarmed by the BEACON center, said Pennock, because it shows how complex systems can evolve through the mechanisms of evolution. They have been "trying to attack the whole project because we can see evolution doing what they've claimed it can't do. That's the thing about observing evolution in action. It's compelling to students because they can see for themselves. It's not just that you're telling them; they can see it."

Avida has been used to do forefront research in evolutionary theory (see, for example, Adami et al., 2000; Lenski et al., 2003; and Yedid et al., 2008). In addition, once an evolution algorithm has been implemented in a computer system, it can be used to solve engineering problems. Pennock noted that evolutionary computation has yielded safer cars, smarter robots, and steadier rockets. This is another way of demonstrating how evolution makes a difference in our lives. "This is something that is very pragmatic," said Pennock. "Evolution works."

These kinds of success stories can be disseminated to the public through a variety of means. A recent USA Science and Engineering Festival featured the practical applications of evolution through the Evolution Thought Trail, a collaborative effort among some 15 disciplinary societies and the National Academy of Sciences.[4] Presentations on influenza viruses, robot controllers, and pest management all have drawn considerable attention. These kinds of outreach efforts "give people a way to start thinking about the process."

Evolution and the Nature of Science

Finally, Pennock observed that, far from being an uncertain science, evolution is science done right. (Box 2-1 describes some of the dimensions of thinking evolutionarily.) As such, it is one of the best examples available to illustrate the nature of science. It illustrates the links between observations and explanations, indirect evidence and experimental results, and causes and effects. "We need to be using evolution to teach about the nature of science," Pennock said.

EVOLUTION IN *MOLECULAR BIOLOGY OF THE CELL*

Over the five editions of *Molecular Biology of the Cell*, cell biologists have grown increasingly aware of the enormous complexity of the chemistry in cells, said Alberts, one of the co-authors of the popular and esteemed textbook. Nearly all cellular processes are driven by groups of 10 to 20 proteins organized as protein machines and incorporating ordered protein movements. Furthermore, these processes occur through elegant mechanisms that themselves are too complex to predict.

Nevertheless, there is a way to shortcut this complexity. Because of evolution, organisms living today have homologies where similar structures or functions were inherited from a common ancestor. For this reason, the shortest path for working out a mechanism in human cells often starts with molecular studies in simpler model organisms. For example, a comparison of genomic sequences for various species of animals shows that the gene that causes cystic fibrosis in humans when it is mutated is very similar across organisms. Many other genomic regions are also care-

[4] See *http://www.ashg.org/education/evolutiontrail.shtml*.

fully conserved over evolutionary time, yet biologists know very little about why many of these regions are conserved or what they do. "When you find these kinds of sequences, what it means is that this thing has some important function, [but] we have no idea what that function is, so it directs what biologists do," said Alberts.

Beyond the Textbook

Textbooks emphasize what scientists have learned, but the most important objective in science education is to teach people what science is, said Alberts. The irrational thinking that is widespread in America "is the strongest argument I can think of for refocusing our education system at all levels on understanding the nature of science, training people how to think rationally, solve problems, and use evidence. Most of them will never be a scientist, but they need that to deal with the world around them."

As John A. Moore emphasized in his *Science as a Way of Knowing* project, it is not enough to tell people about evolution, Alberts observed. They need to understand the nature of science, but that is not happening today. Alberts told an anecdote about a third grader returning from school who told his scientist mother, "Now, I understand science. It's the same as spelling. You just have to memorize it because it does not make any sense." As Alberts said, "I wish every college professor would soak that in because we teach this way even in college science."

Many Americans also mistakenly believe that science is what scientists believe, religion is what religious leaders believe, and both are equivalent dogmatic belief systems. If that is true, according to this line of reasoning people can choose either system. "If you think about how we teach science, this is not such as a surprising conclusion."

As editor of *Science*, Alberts has been working to redefine science education, and the key to this redefinition is the introductory college science class. These classes need to address all four strands of science proficiency described in the publication *Taking Science to School* (National Research Council, 2007):

Strand 1: Know, use, and interpret scientific explanations of the natural world.
Strand 2: Generate and evaluate scientific evidence and explanations.
Strand 3: Understand the nature and development of scientific knowledge.
Strand 4: Participate productively in scientific practices and discourse.

All of the strands are of equal importance in high-quality science education, said Alberts. But only one involves knowing what scientists have discovered about the world. The other three involve how scientists learn about the world. A valuable activity for scientific societies would be to work with other societies and institutions to reshape college introductory biology courses to address all four of these strands, said Alberts.

Scientific Societies

Another valuable role for scientific societies would be to emphasize the importance of high-quality, low-resource lab modules that stress student inquiry to replace the standard, follow-the-instructions, "cookbook" college laboratories. "I was in the laboratory for three years at Harvard, for three afternoons a week," said Alberts. "Basically, I was learning how to cook. I didn't know what science was." In 2011, *Science* conducted a contest for the best inquiry lab modules for introductory college science courses. A module is defined as something that takes 8 to 50 hours of student work, which makes the module small enough to transfer from place to place. The 15 winners will be announced throughout 2012. Once a month (with three months featuring two winners) *Science* will publish a two-page printed article by the originators of the winning module(s), accompanied by on-line supplementary material containing all of the instructions needed to replicate the lab. The contest will be repeated in 2012, with winners being published in 2013.[5]

Finally, Alberts suggested that scientific societies could work with each other and with other organizations to increase the importance and prestige associated with being a great teacher of science. Focus groups have revealed that a failure to understand the nature of science lies at the heart of the evolution versus creationism debate (e.g., Labov and Kline Pope, 2008). "Our teaching of science as the 'revealed truth' has not worked," said Alberts. "It also has not worked to create a population that understands science well enough and can think rationally well enough to confront politicians when they say things about climate change—as they're doing now—that are totally wrong."

Alberts briefly described an introductory college-level biology class at the University of Minnesota. The class takes place in a room with large tables that can seat nine students and have two laptops connected to the Internet. These tables can project what is on either of the screens on an overhead screen, and the teacher can reproduce what is on one screen

[5] For access to these modules and a more extensive description of the initiative, see *http://www.sciencemag.org/site/feature/data/prizes/inquiry/*.

BOX 2-2
Hooking Students with Human Behavior

Marlene Zuk, professor of biology at the University of California, Riverside, works on sexual selection and the evolution of mating behavior. "Sex is a great motivation for people to learn about things, and I'm surprised that no one else has suggested that as a motivation for students."

The biological differences between men and women evolved; indeed, Darwin wrote a whole book titled *The Descent of Man, and Selection in Relation to Sex* (Darwin, 1871). "What makes males different from females is an extremely important evolutionary question that we can answer using the exact same tools that we use to address other scientific questions," Zuk said.

Students inevitably ask questions about human behaviors. But a study of evolution makes it possible to look at the evolution of reproductive behaviors in a wide range of organisms, including humans. The same questions can be asked: On what evidence is a conclusion based? Does a particular caricature of human behavior—such as whether men are more promiscuous than women—have any basis in evolutionary science? "It's not necessarily sidestepping the controversies," said Zuk. "It's giving students the tools to talk about them without assuming that, 'If I go in for this evolution thing, it necessarily means I have to think a certain way about human behavior.'"

on all the screens in the room.[6] "As you might imagine, people who take Biology I this way think completely differently about what science is than do the students who take biology sitting in a big lecture hall, more or less memorizing what the teacher has said." (Box 2-2 describes another way of interesting students in evolution.)

DISCUSSION

Changing Attitudes

In response to a question about whether the 40 percent of high school biology teachers who doubt evolution were science majors or teaching majors, Pennock pointed out that they were all undergraduates at one time, whether they were biology majors or not. Scientists have a tendency to push the blame for not understanding evolution to earlier and earlier ages, whether college, high school, elementary school, or parents' attitudes. But in the end, he argued, "it's our fault." Biology teachers were not

[6] Additional information about the physical facilities and the changes in pedagogy that those facilities have encouraged that are different than what is possible in large lecture halls is available at *http://www.classroom.umn.edu/projects/alc.html.*

shown effective ways to teach evolution when they were undergraduates, so it is not surprising that they struggle as teachers. "If other people aren't doing it, it's because we didn't do a good job when they were our students. So whether they're majors or non-majors, part of what we need to do is clean up our own house." In that respect, the most important audience for how to teach evolution better is current faculty, said Pennock. "If we can't convince them to do this, how are we going to have a hope of convincing anyone else?" Teaching evolution across the curriculum, as well as modeling effective teaching approaches, is a way to break out of this cycle, said Pennock. Effective teachers can show their colleagues how to teach evolution well, and effectiveness will spread.

Elvis Nunez from the University of Florida and Caribbean Examination Council reported on the negative attitudes of high school teachers with whom he has worked. The teachers said things such as "I will start believing in evolution when it starts affecting my life," "Nothing good has come out of evolution," "I can teach and live without knowledge of evolution," and "Even college students have trouble with the subject, so why teach it in high school?"

Edward Egelman from the University of Virginia pointed out that people may not be rejecting science so much as deciding that evolution is somehow controversial within science, in the same way that they have been convinced to think of climate change research as "bad science." Alberts responded that people do not know enough about science to deal with bad information. "They may think that science is wonderful because it brought them the iPhone. But it doesn't mean that they understand enough about it to be able to deal with the modern world of confusing politicians, salesmen, and everybody else trying to get your money or your vote."

Randall Phillis from the University of Massachusetts, Amherst, pointed out that some fundamentalist preachers tell people that if they believe in evolution, they will be damned. This is why some students do not show up when evolution is taught—they fear for their souls, and they will not question their faith.

Biology teacher Paul Strode from Fairview High School in Boulder, Colorado, expressed the view that when students have a belief, whether religious or otherwise, that is in direct conflict with known scientific fact, they should be challenged to reconsider that idea. "We're not challenging those beliefs well enough."

John Staver from Purdue University pointed out that Jesus told his disciples in the Bible to try to understand problems. "The whole notion of religion as memorizing something and not thinking about something and being extremely certain about everything is not what the Rabbi from Nazareth did or advocated," he said. If scientists and religious scholars worked together, they could devise new ways to interact with people who oppose teaching and learning about evolution.

Textbook author Joseph Levine gives a talk during his in-service professional development sessions for teachers about the relationship between religion and science from a personal perspective. He goes back to the first verses of Genesis in Hebrew, translates them for teachers, and works through their meaning. Non-condescending, inclusive messages about how people reconcile faith and science may not change minds, he said, but it can open them.

Maxine Singer reiterated her opening remarks that it is a mistake to challenge a person's faith. "People of great faith have existed since the beginning of human time. Many people depend on that in different ways. . . . Better to try to figure out how to teach biology to people in a way that people will learn if not accept."

Student Motivation

A major theme of several discussion sessions was getting students motivated and emotionally invested in learning about evolution. Caitlin Schrein, a Ph.D. student at Arizona State University, is trying to demonstrate the relevance of evolution to current and personally relevant topics—for example, by teaching about evolution in the context of humans. She also pointed out that the students in an introductory biology course in college have a huge range of backgrounds, from those who have passed AP Biology to those who have never taken biology before. What are the competencies that should be expected of incoming students?

Mark Schwartz from the New York University School of Medicine also observed that it is important to identify real-world benefits of applying evolutionary science, such as understanding the phylogenetics of infectious microorganisms or the metastatic spread of cancer. In addition, Phillis listed invasive species, antibiotic resistance, and the risks of monocultures in agriculture as examples of evolution in action. "Pretty much every day of the semester, we're going to get to the place where we talk about something cool in evolutionary biology."

Schwartz also noted that undergraduates are motivated if they know that something is going to be on the test. "If our goal is to reach everybody, which is one of the carrots that works. It's not the only one, but that is what drives much of undergraduate education. For those of us who teach, the common refrain is, 'Will it be on the test?' And the answer is, 'Yes, you have to understand the concept, not the memorization.'"

Celeste Carter from the National Science Foundation agreed that tests are important to students, but so are teaching styles. Although lecturing can be important for transmitting blocks of information, instructors also can be facilitators of learning rather than directors of learning. "You don't always have to be the person with this font of knowledge that you're going to pour out." For example, an approach that can interest students more than lecturing is problem-based learning, she said. Undergraduate research experiences of many different types also can convince students to remain in science disciplines.

Regarding tests, a strategy Carter used when she taught is to have students take an exam twice. The first time they do it closed book and on their own. The second time they can bring anything they want into the classroom and have an open discussion among themselves about the answers. Many students told Carter that they learned more from discussing their viewpoints than they did from almost any other activity they did in her classroom.

David Mindell from the California Academy of Sciences in San Francisco remarked on the value of getting students into the field to connect evolutionary biology with nature. Students are increasingly from urban populations and settings, he said. "They have no feeling for the organism, in my experience in teaching undergraduates, or they have relatively little. It can make a huge difference to take them out on at least one and ideally multiple trips to the field to let them see organisms in the wild." Such experiences can cement the concepts students learn in a classroom and help them become scientifically literate adults.

According to Richard Potts, Director of the Human Origins Program at the Smithsonian Institution, it is possible to use depictions of evolution in popular culture to teach students, including both realistic ideas about evolution and "terribly wrong" ideas. For example, a discussion of phylogenetic analysis on the CSI television show can motivate students to learn more about the subject. Students "suddenly feel that now it's relevant to them because, well, if it's on CSI, then it's something they care about."

Steve Klein from NSF emphasized the value of knowledge for its own sake. People are very interested in basic sciences such as astronomy, whether it benefits their lives or not. "We need to explain to people that it's a natural human function to try to understand the world we live in."

Better Preparation for Students and Teachers

The summer between high school and college offers many opportunities to motivate and prepare students for college, said Schwartz during his prepared remarks as a panelist. Bridge classes, math boot camps, laboratory classes, mentoring opportunities, seminars on study habits, and many other possibilities exist. The key, said Schwartz, "is to have a repertoire of pedagogies to be able to address as many of those students and as many of those groups as you can, . . . whether you are talking about high school biology, undergraduate biology, community colleges, or medical school."

Carter pointed out during the ensuing discussion that students who take courses at community colleges constitute roughly half of the undergraduates in the United States. These students can range from 12 to 64 years of age and have widely divergent backgrounds. "You have to be creative," she said. "You have to think about and find out who is sitting in that classroom in front of you and then think about strategies that are going to motivate and keep each one of those people engaged."

Improving the subject matter knowledge of teachers was one of the motivations behind the development of UTeach at the University of Texas, said Potts, where undergraduate science majors earn a teaching certificate in four years and are ready to teach high school science when they graduate. Much of the responsibility for the UTeach initiative lies with the university's science faculty.[7] "It's not any surprise that a lot of high school science teachers don't really understand science because they're not science majors in large part, but that's beginning to change," Potts noted.

During the general discussion, Schrein briefly summarized a survey on science education in elementary schools of 1,100 teachers, principals, and district administrators at 300 California public schools (Dorph et al., 2011). Only 10 percent of elementary students regularly experience hands-on science practices, according to the survey. The obstacles reported by teachers, principals, and administrators to teaching science include the lack of funds for supplies, not enough time, and insufficient teachers' training. According to the survey, 40 percent of elementary teachers spend fewer than 60 minutes teaching science per week.

Jay Labov of the National Academy of Sciences and the National Research Council, and study director for the NRC report that resulted in the current restructuring of several Advanced Placement science courses, said that the new AP Biology program has the potential to be a game changer. (See Chapter 6 for a description of these changes.) People will be less likely to attack the restructured course's increased emphasis on evolution as a "big idea" and "unifying theme" because AP Biology offers too many benefits in terms of college admissions and credit. Students who take the class may not come to accept evolution, but they will at least learn about the subject. He also emphasized the influence of AP courses on other parts of the high school curriculum as well as on middle schools and postsecondary education.

Finally, Maxine Singer and several other people pointed to the impor-

[7] Additional information about UTeach is available at *http://uteach.utexas.edu*. This model has been promulgated through the National Mathematics and Science Initiative (*http://nationalmathandscience.org*) and is now available at universities across the United States.

tance of reaching students who do not take AP Biology. These students will constitute the large majority of the general public in the future, and their understanding of evolution will dictate which attitudes are most prevalent.

REFERENCES

Adami, C., Ofria, C., and Collier, T. C. 2000. Evolution of biological complexity. *Proceedings of the National Academy of Sciences* 97(9):4463-4468.

Berkman, M. B., and Plutzer, E. 2011. Defeating creationism in the courtroom, but not in the classroom. *Science* 331(6016):404-405.

Darwin. C. 1871. *The Descent of Man, and Selection in Relation to Sex*. London: John Murray.

Dorph, R., Shields, P., Tiffany-Morales, J., Hartry, A., and McCaffrey, T. 2011. *High Hopes—Few Opportunities: The Status of Elementary Science Education in California*. Sacramento, CA: The Center for the Future of Teaching and Learning at WestEd.

Labov, J. B., and Kline Pope, B. 2008. From the National Academies: Understanding our audiences: The design and evolution of science, evolution, and creationism. *CBE/Life Sciences Education* 7:20-24.

Lenski, R. E., Ofria, C., Pennock, R. T., and Adami, C. 2003. The evolutionary origin of complex features. *Nature* 423:139-145.

Miller, K. R., and Levine, J. S. 2004. *Biology*. Upper Saddle River, NJ: Prentice-Hall.

Miller, J. D., Scott, E. C., and Okamoto, S. 2006. Public acceptance of evolution. *Science* 313: 765-766.

National Research Council. 2007. *Taking Science to School: Learning and Teaching Science in Grades K-8*. Washington, DC: The National Academies Press.

Yedid, G., Ofria, C. A., and Lenski, R. E. 2008. Historical and contingent factors affect re-evolution of a complex feature lost during mass extinction in communities of digital organisms. *Journal of Evolutionary Biology* 21(5):1335-1357.

3

Learning About Evolution: The Evidence Base

Biological evolution is a difficult concept to learn, as several people at the convocation emphasized. It involves complex biological mechanisms and time periods far beyond human experience. Even when students have finished a high school or college biology course, there is much more to learn about the subject.

The difficulty of teaching evolution both complicates and invigorates research on evolution education. To present what is known and not known about the teaching and learning of evolution—which is a standard feature of convening events organized by the Academies—Ross Nehm, associate professor of science education at Ohio State University, gave an overview of the research literature on evolution education and then talked in more detail about his own research.

THE EVIDENCE BASE

The literature on teaching and learning about evolution is extensive. In 2006 Nehm reviewed 200 of more than 750 papers published thus far about evolution education, identifying both strengths and limitations of the approaches taken in those studies (Nehm, 2006). This literature demonstrates that the general public, high school students, undergraduates, biology majors, science teachers, and medical students all have low levels of knowledge and many misconceptions about evolution (Nehm and Schonfeld, 2007). Furthermore, as with other areas of science, many of the same misconceptions persist in all of these populations. "They don't go away," said Nehm. "Whatever instruction is happening at early levels, it's not ameliorating the problems that we have."

In education, the only way to make robust causal claims is through a randomized controlled trial (RCT), but no such trials have been conducted for evolution education. "If you want to make causal claims, there is no causal literature to refer to."

Fortunately, other research tools can be used with educational interventions to draw conclusions that can guide policy. A group receiving an intervention can be compared with a group not receiving the intervention. Interventions can be done without a comparison group—for example, by looking at a group before and after an intervention. Survey research can yield associations, although survey research cannot determine whether these associations are causal. Finally, case studies, interviews, and other forms of qualitative research can reveal new variables and possible associations.

Nehm's 2006 review of the literature found no intervention studies with randomized control groups, 6 intervention studies with comparison groups, and 24 other studies that employed various intervention techniques. Also, some of the interventions were quite brief—just one to three weeks—a period during which substantial changes are unlikely to occur, given the difficulties of teaching evolution. One conclusion is obvious, Nehm said: "We need to do some randomized controlled trials to see what works causally in terms of evolution education."

Nehm also pointed out that documenting learning outcomes is critically important in education research. According to the report *Knowing What Students Know: The Science and Design of Educational Assessment* (National Research Council, 2001), "assessments need to examine how well students engage in communicative practices appropriate to a domain of knowledge and skill, what they understand about those practices, and how well they use the tools appropriate to that domain." Yet most tests today, including those that dominate biology curricula, assess isolated knowledge fragments using multiple choice tests. Students may be learning about evolution, "but if we can't measure that progress, we can't show that what we're doing has any positive effect. So we need assessments that can measure the way people actually think."

The problems caused by inadequate metrics are particularly obvious in the literature on teacher knowledge of evolution, Nehm said. Only five intervention studies exist, and three of them assess teacher's knowledge of evolution using a multiple choice or Likert scale test (Baldwin et al., 2012). This lack of careful metrics "is really concerning," said Nehm. Evolution assessments must be developed that meet quality control standards established by the educational measurement community, or robust claims, causal or otherwise, cannot be made.

In summary, research has established key variables that should be investigated and many possible beneficial interventions. But the research literature on evolution education lacks robust, causal, generalizable claims relating to particular pedagogical strategies and interventions. It also lacks measurement instruments that meet basic quality control standards and capture authentic disciplinary practices. Finally, the research lacks consistent application of measurement instruments across different populations. "This is a call to action," said Nehm. "We need to gather and do a national randomized controlled trial of some of the most likely and agreed upon variables and test their causal impact on students' learning of evolution."

NOVICE TO EXPERT REASONING

In his own research, Nehm and his colleagues have been studying how different groups, from novice to expert, think about problems.[1] Using performance-based measures in which research participants are asked to solve evolutionary problems, they have looked at 400 people—including non-majors who have completed an introductory biology course, students who have completed a course in evolution, students who have completed an evolution course as well as more advanced coursework, and a group of biology Ph.D. students, assistant professors, associate professors, and full professors (Nehm and Ha, in preparation).

The study measured people's ability to explain evolutionary change across a variety of contexts, not through multiple choice questions. In general, this technique revealed many more gaps in evolutionary understanding than would simpler assessments. For example, students have a harder time explaining evolutionary change (in writing or orally) than recognizing accurate scientific elements of an explanation when presented in a multiple choice test (Nehm and Schonfeld, 2008). Or, as Nehm put it, knowing the parts and tools needed to assemble furniture does not mean that you can build it. Students may have a lot of knowledge about evolution but not be able to use that knowledge to create a functional explanation. "This is a tough competency," explained Nehm. "If you asked any of your students, and I encourage you to do this, 'Can you explain how evolutionary change occurs?' you will be startled at their inability to articulate their understanding because they are never asked to do that."

In addition, people have a tendency to mix naïve and scientific information together in their explanations. Naïve ideas include, for example, the notions that the needs of an organism drive evolutionary change or

[1] A summary of the general research on differences between novices and experts can be found in National Research Council (2000).

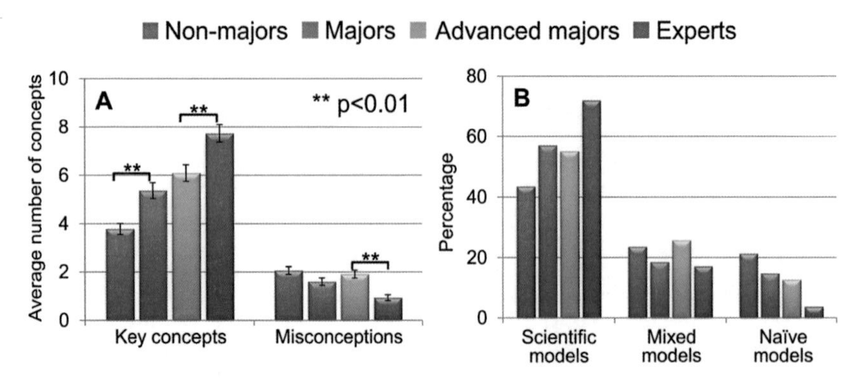

FIGURE 3-1 Misconceptions decrease with educational level but never entirely disappear (left), while mixed models of evolutionary change remain as common in advanced biology majors as in non-majors (right). The vertical scale on the left measures the numbers of key concepts and misconceptions used by respondents, while the vertical scale on the right measures the percentage of respondents using different kinds of explanatory models. SOURCE: Nehm and Ha, in preparation.

that putting pressure on animals will cause them to evolve. The mixing of naïve and scientific ideas is difficult to measure with multiple choice tests, but open response explanations can reveal the relative contributions of each category of information.

Of 428 people—107 from each group—the experts (the combined group of Ph.D. students, assistant professors, associate professors, and full professors) knew more key concepts and had fewer misconceptions (Figure 3-1). Some Ph.D. students still have naïve ideas about evolution, and occasionally a professor, although that was uncommon. People learn more about evolution as they take more courses, but a surprising number do not get rid of their misconceptions.

Moreover, as shown in Figure 3-1, up to 25 percent of the advanced majors who have taken an evolution course and other advanced courses still construct mixed models of evolutionary explanations that combine naïve and scientific ideas. The use of exclusively scientific models increases with educational level, but this use never gets above 60 percent of students in Nehm's research. Furthermore, many students have only naïve ideas, although this percentage declines with educational level.

SEEING BENEATH THE SURFACE

Research shows that novices tend to get tripped up by surface features of problems, such as the context, format, or details of a problem, rather than grasping a problem's underlying structure. They think that similar problems framed in different ways are actually different problems, whereas experts see the similarities that are not apparent on the surface of those problems. Nehm and Ridgway (2011) applied this analysis to evolutionary biologists and to non-majors who had taken an introductory biology course and found the same thing. For example, elements of natural selection were linked consistently by experts but haphazardly by novices.

An especially intriguing finding was that students tend to draw on different misconceptions in trying to solve different types of problems. In problems involving plants, animals, or bacteria, for example, their misconceptions tend to differ based on the type of organism they are being asked to consider. Thus, in teaching evolution, the types of misconceptions that teachers think they are tackling are correlated with the kind of problem they give students. For example, even though teachers may think they are describing a problem involving natural selection, a particular surface feature of the problem may keep students from recognizing the connection. Likewise, in assessments, the type of misconception being assessed is correlated with the type of problem students are trying to solve. "That's a big implication," said Nehm. Students see natural selection in a cheetah and in a bacterium as completely different processes. This is one way in which they glue new information onto preexisting naïve ideas.

The importance of surface features has received almost no attention in evolution education, Nehm observed. Introductory biology textbooks use a variety of contexts but never alert students that bacterial resistance to antibiotics is no different from the many other examples of natural selection being described. "We never help students see those parallels." Even after completing an evolution course, only 50 percent of students have "expert-like" perceptions of evolutionary problems. As students progress through biology, their courses do little to help them reason across cases.

Novices' Thinking About Evolution

What are the problems that novices have in thinking about evolution? People have many different kinds of knowledge, including conceptual resources, analytical resources, and factual resources. According to traditional models of problem solving, people draw information from these different kinds of resources and put it in working memory to tackle problems.

Nehm has been testing this concept for evolution. In one experiment, more than 200 participants solved problems in which just one feature was manipulated at a time (Nehm and Ha, 2011). The experiment looked at which surface features are problematic for learners, such as scale (such as intraspecific or interspecific), polarity (such as trait gain versus trait loss), taxon (such as plant or animal), and familiarity. The experiment measured the accuracy of their scientific thinking.

The results of this experiment show that students have more trouble reasoning about the loss of traits than the gain of traits. They also have greater difficulty reasoning about the loss of traits between species than within species. But in reasoning about the gain of traits, there is no difference between the interspecific and intraspecific situations. Similarly, students have more misconceptions about the loss and gain of traits between species than within species. The hardest problem for students to solve, said Nehm, is the loss of traits between species. "If students can handle that, that's the highest level of competency. But do we ask those questions? No."

Also, students use more key concepts in solving problems involving familiar animals than unfamiliar animals, but this trend is not seen for plants, all of which seem to strike students as unfamiliar.

These surface features have a remarkably powerful influence, said Nehm. "If you want to show your class is doing great, I can design an assessment for you. If you want to show your students are failing, I can design an assessment for you. All I have to do is manipulate surface features because students' reasoning is so tied to these features. And yet we pay no attention to this in any textbook or in any assessment."

The bottom line is that "surface features matter, and we need to be more precise in our instructional strategies to deal with these." Because misconceptions are surface-feature specific, instructional examples must be carefully chosen. Furthermore, assessments of competency must include authentic production tasks, such as explaining how evolutionary change occurs, not just fragmented knowledge selection tasks.

EVOLUTION ACROSS THE CURRICULUM

In 2007, Nehm reported on an introductory biology course that was changed so that every topic included evolution, while a parallel course was taught using a traditional curriculum (Nehm and Reilly, 2007). The outcomes were not substantially different. "It's an awful downer at this conference," he admitted.

However, one single study is not enough to draw broad conclusions. For one thing, students have difficulty learning evolution, so teaching it in the same way is probably not going to lead to progress. "If you have a problem with A and you give lots more A, the chances are it's not going to lead to a substantial improvement."

Also, as students work through the biology curriculum, they move from naïve models to mixed models to scientific models, but progress is very slow—25 percent of students who have completed a course on evolution and additional coursework still used mixed models.

Determining the conditions under which students can effectively learn about evolution will require truly randomized controlled trials, Nehm concluded. Developing such trials will be difficult, he acknowledged. "But my perspective—which, again, is only my personal perspective and may be wrong—is that if people can do it in medicine, where people are dying, we should be able to do it in education."

REFERENCES

Baldwin, B. C., Ha, M., and Nehm, R. H. 2012. The Impact of a Science Teacher Professional Development Program on Evolution Knowledge, Misconceptions, and Acceptance. *Proceedings of the National Association for Research in Science Teaching (NARST) Annual Conference*, Indianapolis, IN, March 25-March 28.

National Research Council. 2000. *How People Learn: Brain, Mind, Experience, and School: Expanded Edition*.Washington, DC: National Academy Press.

National Research Council. 2001. *Knowing What Students Know: The Science and Design of Educational Assessment*. Washington, DC: National Academy Press.

Nehm, R. H. 2006. Faith-based evolution education? *Bioscience* 56(8):638-639.

Nehm, R. H. 2007. Teaching evolution and the nature of science. *Focus on Microbiology Education* 13(3):5-9.

Nehm, R. H., and Ha, M. 2011. Item feature effects in evolution assessment. *Journal of Research in Science Teaching* 48(3):237-256.

Nehm, R. H., and Reilly, L. 2007. Biology majors' knowledge and misconceptions of natural selection. *Bioscience* 57(3):263-272.

Nehm, R. H., and Ridgway, J. 2011. What do experts and novices "see" in evolutionary problems? *Evolution: Education and Outreach* 4:666-679.

Nehm, R. H., and Schonfeld, I. 2007. Does increasing biology teacher knowledge about evolution and the nature of science lead to greater advocacy for teaching evolution in schools? *Journal of Science Teacher Education* 18(5):699-723.

Nehm, R. H., and Schonfeld, I. 2008. Measuring knowledge of natural selection: a Aomparison of the CINS, and open-response instrument, and oral interview. *Journal of Research in Science Teaching* 45(10):1131-1160.

4

Confronting Controversy

Three speakers at the convocation specifically addressed the difficulties teachers can face in the classroom when they teach evolution. None had a way to avoid controversy, but all had ways to deal with it.

OVERCOMING FEAR

Paul Strode, who teaches biology at Fairview High School in Boulder, Colorado, grew up in Indiana and went to a small liberal arts college, where he took courses in zoology, genetics, and ecology but learned very little about evolution. After he moved to Seattle to teach high school biology, he left the evolution chapter to the end of the year, as many teachers do, and warned his students the day before the session began that the class was going to discuss evolution because it was part of the curriculum. "I was frightened and nervous, and sure enough the next day when I came in to the classroom there were pamphlets on my desk about things that I couldn't answer, because I had no way to answer them." He went ahead and taught the unit, but "for the next seven years I avoided evolution completely. That scared the heck out of me."

Many new teachers fear that their students will discover that they do not know everything, said Strode. "I didn't want anyone in my classroom, I didn't want the principal there, and I didn't want another teacher to watch me teach, because I thought I might be found out."

After eight years of teaching, Strode went to the University of Illinois for a Ph.D. in ecology and environmental science. "That's when it was revealed to me that I had no idea how science worked." One day one of his thesis advisers asked him what hypothesis he was testing with his research. Strode gave her an answer, and she replied, "No. Those are your predictions. What are your hypotheses?" He reformulated his answer, and she said, "No, you're still giving me predictions. You don't know what hypotheses are, do you?"

"That was almost the deal breaker for me," Strode recalled. "I almost walked home and quit because I thought, 'What am I doing—and what have I done for eight years as a high school teacher—having no idea that science is a hypothesis-based form of inquiry?'"

Strode finished his doctorate and moved to Boulder to teach high school with a whole new outlook. He realized that many of the lab activities he had done in his first eight years of teaching were canned experiments where the outcome was obvious; if students did not get the right answer, they were worried. He started designing activities where the data were messy and the outcome was unknown. He realized that his students were smart enough to learn about statistics, so he taught them about confidence intervals and the analysis of variance.

He also was asked to co-author a book called *Why Evolution Works and Creationism Fails* (Young and Strode, 2009). "That forced me to realize that I wasn't doing as good a job as I could teaching evolutionary theory in the classroom." He began teaching evolution every day so that the subject was woven throughout the curriculum and was not confined to a single unit at the end.

To avoid the problems he experienced, said Strode, teachers need preparation courses on both evolutionary science and on avoiding denialism, whether the subject is evolution, climate change, vaccination, or any other controversial subject. They also need to understand how science works and how that relates to the teaching of evolution. The solution, he said, starts "with kids understanding how science works and teachers understanding how science works and teaching teachers in a more effective way."

"BELIEVING" IN EVOLUTION

Betty Carvellas, who taught science for 39 years at the middle and high school levels and who served as a member of the authoring committee of *Science, Evolution, and Creationism* (National Academy of Sciences and Institute of Medicine, 2008), taught her students that they do not necessarily need to "believe" in evolution, but they do need to understand the scientific evidence demonstrating that evolution is a fact. She would not have been comfortable saying this early in her career, but when she did become comfortable doing so, it allowed her to do two things. It informed her students that evolution would be a part of every experiment they did all year long. It also gave them the right to say, "I don't believe in evolution," because, as Carvellas said, "I don't believe in evolution either. A belief is one thing, and a scientific fact is something altogether different."

Evolutionary thinking is not possible without scientific thinking, she said. If students do not understand the nature of science, the processes of science, and the limitations of science, they are not going to understand evolution. These ideas about scientific thinking have to be built starting in kindergarten. Young students love to do the things that develop scientific understanding, such as asking questions, developing models, planning and carrying out investigations, analyzing and interpreting data, thinking mathematically, constructing explanations, and engaging in argument from evidence. "If they haven't had that opportunity [by high school], they're fearful of this because they don't know what the right answer is."

Teachers in earlier grades, starting in kindergarten, do not need to mention evolution, Carvellas said, but they must introduce the concepts that will make an understanding of evolution possible. This will require more professional development and resources for teachers. For example, the new frameworks for science education call for students at the end of second grade to understand that some of the plants and animals that once lived on earth are no longer found anywhere, although others now living resemble them in some ways (National Research Council, 2011). Second graders are not ready to understand common ancestry, but later in their schooling they will have an easier time understanding the concept. "If you plant those seeds and let kids work with them, it's going to make our lives so much easier when they get to high school and college."

Carvellas built evolution into her entire course, whether the subject was ecology, environmental change, genetics, or any other subject. By the time her class studied evolution directly, they had a basis for how it happens, they were more motivated, and they were more interested.

Many teachers in some parts of the country cannot deal with the conflicts with parents and students who confront them on a daily basis, so they avoid evolution. One teacher told Carvellas that her students "would challenge her on day one, and if they found out that she accepted evolution, they would make her life miserable for the entire school year."

But Carvellas also said that students are quite receptive to being told that what their friends told them is wrong. "They love finding out that [a misconception] isn't true and here is why we know it's not true and here's what really works. . . . They love knowing more than their friends know."

KEEPING AN OPEN MIND

David Hillis, Alfred W. Roark Centennial Professor in Natural Sciences at the University of Texas and a member of the National Academy of Sciences, said that he has been teaching biology for 30 years and always in states where evolution is controversial, including Kansas, Florida, and Texas. At the beginning of his introductory biology courses in college, or even his advanced evolution courses, students often tell him that they have a religious problem with evolution. Rather than confronting them, Hillis asks them to try to keep an open mind, listen to what he has to say about evolution, and then come back to him if they still have problems. "In 30 years, I've never had a student come back to me. I've never had a comment on an evaluation complaining about evolution." People who have religious objections to evolution largely do not know what evolution is. Many of their objections can be overcome by "simply addressing that ignorance."

In teaching evolution, Hillis starts with familiar examples from the present and recent past and gradually works his way toward the distant past. "They can see that the exact same concepts and things that they know and can understand in the present or in the recent past apply to the ancient past."

He also seeks to show how the mechanisms of evolution that can be observed today are sufficient to account for major evolutionary changes over long periods of time. Students need to grasp the deep time of Earth to understand why these mechanisms have had enough time to work. "They have a hard time understanding the difference between a thousand and a million, much less between a million and a billion. Once they have used mechanisms to help them understand the depth of time we're talking about, and you start multiplying the kind of changes we see over short time to those longer times, they begin to understand how this can all work."

Students need to understand that evolution has practical applications by learning about examples of evolution in action. They need to be shown applications of evolution in human health, agriculture, industry, and basic science.

Instructors also need to demonstrate that evolution is an experimental and an observational science, said Hillis. Few biology courses or textbooks emphasize the point that all the basic mechanisms of evolution can be observed directly and confirmed experimentally, and classrooms should feature these demonstrations and experiments to a much greater extent.

Finally, as emphasized throughout the convocation, evolution needs to be applied in every unit of biology courses and in every chapter of biology textbooks. Textbooks still need to have chapters on evolution to explain the details of the process, said Hillis, just as advanced courses on evolution will still be necessary. But evolution should not be limited to those sessions or classes.

REFERENCES

National Academy of Sciences and Institute of Medicine. 2008. *Science, Evolution, and Creationism*. Washington, DC: The National Academies Press.
National Research Council. 2011. *A Framework for K-12 Science Education: Practices, Crosscutting Concepts, and Core Ideas*. Washington, DC: The National Academies Press.
Young, M., and Strode, P. K. 2009. *Why Evolution Works and Creationism Fails*. Piscataway, NJ: Rutgers University Press.

5

Broadening the Target Audiences

One session of the convocation was devoted specifically to consideration of the intended audiences of evolution education. High school and college students are of course a major audience, but many other audiences were mentioned, from preschool children to legislators. And for all of these audiences, including students, how to deal with opposition to the teaching of evolution is a major consideration.

STARTING YOUNG

Richard Potts from the Smithsonian Institution's Human Origins Program of the National Museum of Natural History reported on an informal survey conducted by a colleague of where ninth grade biology students hear about evolution. Number one was family and friends; number two was church; number three was television; and number four was school and science classes. These results suggest that people develop an understanding, or a lack of understanding, of evolution from many different sources, Potts said. Thus, evolution education needs to articulate with messages and information for many other audiences, from church groups to the broad public.

John Staver from Purdue University said that understanding starts in infancy. From an early age, many young people in the United States absorb negative views about evolution. Turning this situation around requires talking about more than science; it requires talking about religion. "The most important factor in learning anything new is what the learner already knows. And in many situations, the learner already thinks that she or he knows that evolution is evil and that you're going to go to hell in a hand basket if you believe in it."

Debra Felix from the Howard Hughes Medical Institute agreed that "by college, it's far too late." Children need to start learning important concepts even before they enter school. "Three- and four-year-olds are extremely curious and extremely capable, and we waste those years by not trying to teach them some of these things." In part, this means reaching out to parents.

POTENTIAL AUDIENCES

Allen Rodrigo, the director of the National Evolutionary Synthesis Center (NESCent, which is described in Chapter 6), discussed some of the many audiences that NESCent is trying to reach. It has a program in evolutionary medicine, a K-12 outreach initiative for minorities who are underrepresented in science, a Darwin Day program, a road show that goes to rural communities, and an ambassador program that extends overseas. "These are constituencies that we feel are important, but we've developed this with an almost intuitive gut instinct that these things are going to be important." As Ross Nehm's argued (see Chapter 4), an important question is how to measure the effects of these programs and any trickle-down effects they have on other groups, Rodrigo observed.

Other important target audiences are parent-teacher associations, boards of education, park rangers, and boy and girl scouts troops. Particularly influential groups include advertisers, public relations firms, entertainers, and game designers. For example, the National Academy of Sciences has an office in Los Angeles called the Science and Entertainment Exchange[1] that works with entertainment industry professionals in Hollywood to help them better understand science in the context of television shows and movies.

Rodrigo noted that journalists are another important audience. Sessions for editors, producers, and reporters could introduce them to the issues and show them how omnipresent and important evolution is in everyday life. A more diverse audience is the group of people who use social networking. The conversations occurring over these networks could be leveraged to have a broad impact. Blogs, short films on YouTube, science cafes, and other forms of new media, and especially social media could all be used more effectively to convey information about evolution.

An important model for outreach and communication is the work done by Michael Zimmerman, who has been convening the Clergy Letter

---

[1] Additional information is available at *http://www.scienceandentertainmentexchange.org*.

Project and promotes *Evolution Weekend*, which provides an opportunity for congregations to discuss the relation between science and religion on the Sunday in February closest to Darwin's birthday.[2]

DEALING WITH OPPOSITION TO EVOLUTION

Many people have been exposed to very negative ideas about evolution, said Nancy Moran, William H. Fleming Professor of Biology at Yale University, a member of the National Academy of Sciences, and a member of the organizing committee for the convocation. In talking with such people, "the worst thing to do is immediately draw a line in the sand and start talking about evolution versus religion," said Moran. "Immediately they'll clam up and feel that somehow they're doing the wrong thing. Many of them have deep-seated feelings that they're doing wrong by learning this."

One productive way to engage in such a conversation is to get people interested in the science—in mutations, alleles, how genetic variants spread in populations, how they contribute to human disease. "You have to go around them rather than confront them directly," Moran said. She added that it is useful to cover some of the scientific controversies in evolutionary biology where biologists currently disagree. That allows people to see that "it's not a big conspiracy. . . . When they see that, they trust the science more."

In contrast to some of the statements offered by other participants, Connie Bertka, former Director of the Dialogue on Science Ethics and Religion at the American Association for the Advancement of Science who now teaches a course at Wesley Theological Seminary on science and religion for students studying to be ministers, observed that people inevitably bring their worldviews to discussions of evolution, but religious worldviews are not necessarily a problem. "There are actually a lot of people in religious communities who are eager to incorporate what science has learned about the world into their theologies. The scientific community ought to be looking at the ways to do everything we can to help . . . because in the long run the message has to come from within those communities. We can't come from the outside and tell people how to reconcile what they see as conflicts, but we can support people within those communities who are trying to do that."

From a Christian perspective, said Bertka, people who grapple with these questions are "doing the same thing that Christians have done throughout time." Christians continually have had to struggle with the

[2] Additional information is available at *http://theclergyletterproject.org/rel_evolution_weekend_2012.htm.*

tenets of their faith in light of new scientific knowledge and understanding about the natural world. Religious traditions change over time, said Bertka, and science needs to engage with this change. "There's no magic bullet here."

Carol Aschenbrener from the Association of American Medical Colleges stated that educators need to help parents see why it is important for their children to understand evolution. "There have to be some concrete and very pragmatic examples of why it's in their best interest and in their children's best interest to understand that." She said that she was the product of a parochial education, yet she studied evolution every single year after the fourth grade. "It was not a contradiction. It was an important part of understanding the complexity of creation."

As Ida Chow from the Society for Developmental Biology and a member of the organizing committee said, "The majority of the people in the country are reasonable. They just don't understand. Here is the opportunity for us to talk to them in a non-threatening way and explain what evolution is and make it relevant to their lives. . . . It is not an easy task, but I think we can all do it if we put our hearts and minds to it."

6

Progress and Resources

Many participants at the convocation described both the progress that has been made in the past implementing evolution across the curriculum and available resources that can enable greatly accelerated progress. Educators have had particular success reforming Advanced Placement (AP) Biology and some aspects of premedical and medical education, as described in the first part of this chapter. Professional societies also can have a significant impact on education at all levels, as four representatives of those societies observed. In addition, the resources available to make progress in teaching evolution across the curriculum—a few examples of which are described in this chapter—are continually expanding.

CURRICULUM REFORM INITIATIVES

The Biological Sciences Curriculum Study

The idea of teaching evolution as a major theme in biology is not new, observed Paul Beardsley, formerly a Science Educator at the Biological Sciences Curriculum Study (BSCS), now at California State Polytechnic University, Pomona, and also a member of the organizing committee for the convocation. Curriculum development at BSCS in the 1950s and 1960s was based on nine major themes, including evolution, "diversity and unity," and "science as inquiry." What has changed since the 1960s, said Beardsley, is that educators have learned how difficult it is to teach these concepts. "To me, evolution is the most difficult set of concepts to teach in all of introductory science."

Beardsley cited several lessons he has drawn from experience and research that need to be taken into account when designing a curriculum to teach evolution. First, people come to class with pre- and misconceptions about how the world works, and these misconceptions need to be recognized by faculty and addressed in curriculum materials. Second, students need to develop a deep factual understanding based on a conceptual framework grounded in evolutionary science. Third, students need practice thinking about their own learning, which cognitive science researchers call metacognition. Finally, students need a source of motivation, particularly those who are underrepresented in the sciences. "These are not novel ideas," said Beardsley, "but they need to be a part of our curriculum."

BSCS has completed a project funded by the National Institutes of Health (NIH) to develop a rigorous contemporary evolution and medicine curriculum based on inquiry, constructivism, and relevance to students' lives. The resulting evolution and medicine curriculum, which was funded by 11 different offices and centers at NIH, is based on the idea that modern health research requires an understanding of evolution. One lesson, for example, discusses the evolution of lactose tolerance in evolution. Students explore data on lactose tolerance and intolerance and develop explanations of the observed global patterns. They can examine mutations that are common in different parts of the world, argue over alternate explanations, and arrive at conclusions about the persistence of lactase, the enzyme which breaks down the sugar lactose, into adulthood in some human populations. In another lesson, students compare genetic sequences across species for a gene associated with cleft palate, explain the results in terms of common ancestry, and explain how natural selection conserved certain sequences of DNA. In another, they use evolutionary principles and concepts to understand influenza by aligning DNA sequences of the hemagglutinin gene and relate the principles of natural selection to the need for new vaccines.

BSCS also has recently finished a major revision of its comprehensive high school biology textbook, *BSCS Biology: A Human Approach*. The intent of the revision has been to help students identify preconceptions, foster metacognitive habits, and build interest through relevant, exciting, and engaging examples.

AP Biology

More than 200,000 high school students take AP Biology every year, with approximately 160,000 to 180,000 sitting for the AP Biology exam. In recent years, the course has been redesigned along the lines recommended in the report *Learning and Understanding: Improving Advanced Study of Mathematics and Science in U.S. High Schools* (National Research Council, 2002). The new course is organized around 4 big ideas, 7 scientific practices, and 17 enduring understandings (see Box 6-1). "It's not about covering 1,500 pages of your favorite textbook," said Spencer Benson, director of the Center for Teaching Excellence and associate professor in the Department of Cell Biology and Molecular Genetics at the University of Maryland, College Park, who was co-chair of the AP biology curriculum redesign committee. "It's about developing a framework to understand all of biology."

The big ideas of the redesigned course emphasize concepts, evidence, and data analysis rather than requiring students to memorize endless facts. Benson also called attention to the practice of connecting and relating knowledge across various scales, concepts, and representations in and across domains. That means looking at evolution from many different biological perspectives and reiterating the idea that evolution is a central component of biology throughout the curriculum.

As a high-stakes exam taken by many students every year, restructuring the AP Biology exam is also "critical," Benson said. The exam is being redesigned to emphasize the concepts, content, and practices that serve as organizing principles for the new curriculum. "People are writing by evidence-based design," said Benson, "which means that every question is linked directly into the curriculum framework and into scientific practices and enduring understandings." Results on the exam will be analyzed to determine whether the new exam is working better than the previous one.

Undergraduate Biology Education

In 2007 the American Association for the Advancement of Science, with support from the National Science Foundation, the Howard Hughes Medical Institute, and the National Institutes of Health, launched a major initiative to develop a shared vision for undergraduate biology education and the changes needed to achieve that vision. As Celeste Carter, a program director in the Division of Undergraduate Education at the National Science Foundation, observed at the convocation, the driving force of the initiative was, "how do you make the biology that we teach as exciting as the biology that we do in our laboratories?" Over the course of the two years, the group held a series of regional meetings and then a national conference with faculty, administrators, representatives of professional societies, and students and postdoctoral fellows. This meeting resulted in the report *Vision and Change in Undergraduate Biology Education: A Call to Action* (Brewer and Smith, 2011), which, as stated in the preface of that

BOX 6-1
Framework for the New AP Biology

The redesigned AP Biology course is organized around four big ideas:

- The process of evolution drives the diversity and unity of life.
- Biological systems utilize free energy and molecular building blocks to grow, to reproduce, and to maintain dynamic homeostasis.
- Living systems store, retrieve, transmit, and respond to information essential to life processes.
- Biological systems interact, and these systems and their interactions possess complex properties.

The course also emphasizes seven scientific practices (enduring understandings), all of which have a connection to evolutionary understanding.

- The student can use representations and models to communicate scientific phenomena and solve scientific problems.
- The student can use mathematics appropriately.
- The student can engage in scientific questioning to extend thinking or to guide investigations within the context of the AP course.
- The student can plan and implement data collection strategies appropriate to a particular scientific question.
- The student can perform data analysis and evaluation of evidence.
- The student can work with scientific explanations and theories.
- The student is able to transfer knowledge across various scales, concepts, and representations in and across domains.

Here is an example of how questions on the AP Biology examination are likely to change.[a] The first question is typical of factual recall questions that are prevalent in current AP Biology tests. The second represents a higher-level question that requires students to demonstrate a greater level of understanding and synthesis of the concept being tested. Graphs next to each question represent the percentage of students who selected each answer:

---

[a] Source: Copyright © 2012 The College Board. Reproduced with permission. *http://apcentral.collegeboard.com.*

AP

Knowledge Skills Behavior Awareness

Try a college-level biology question

The creeping horizontal and subterranean stems of ferns are referred to as:

A. Prothalli

B. Fronds

C. Stipes

D. Roots

E. Rhizomes

A	B	C	D	E	Omit
3%	20%	8%	17%	22%	30%

CollegeBoard

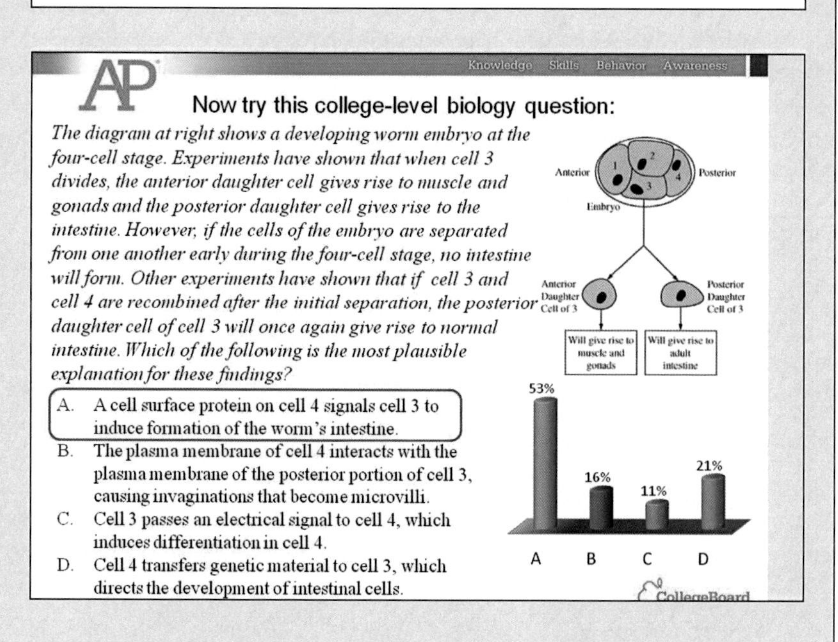

report, "represents the collective wisdom of hundreds of leading life scientists who contributed to the conversations."[1]

The report calls for undergraduates to master five basic biological concepts, the first of which is "the diversity of life evolved over time by processes of mutation, selection, and genetic change." The report explicitly recognizes that evolution "is a thread that should extend all the way through the undergraduate curriculum," said Carter. As the report says, "Because the theory is the fundamental organizing principle over the entire range of biological phenomena, it is difficult to imagine teaching biology of any kind without introducing Darwin's profound idea."

The organizers of the *Vision and Change* initiative are continuing to work to implement the ideas contained in the report, said Carter. A particular need is for funders to decide on levers that they can use to incentivize change.

Premedical and Medical Education

The same year that the *Vision and Change* initiative got under way, the Howard Hughes Medical Institute formed a partnership with the Association of American Medical Colleges to examine the education of future physicians. The report emerging from that partnership identified the most important scientific competencies required of students graduating from college prior to matriculating into medical school as well as the scientific competencies required of medical school graduates as they enter postgraduate training. One of the eight competencies identified as essential for premedical students is that they "demonstrate an understanding of how the organizing principle of evolution by natural selection explains the diversity of life on earth."

The focus on *competencies rather than courses* has several beneficial consequences, said William Galey, director of graduate and medical education programs at the Howard Hughes Medical Institute, during his prepared remarks as a panelist. It provides room in the curriculum for new areas of science and mathematics that need to be addressed. Also, it "liberates the undergraduate curriculum from the tyranny of premed requirements," said Galey. The report has been adopted in principle by the committee that reviews the content and structure of the Medical College Admission Test (MCAT), which will further shift the emphasis in premedical and medical education toward competencies and away from specific courses. A new version of the MCAT is being developed for release in 2015.[2]

---

[1] Additional information is available at *http://visionandchange.org*.

[2] Additional information is available at *https://www.aamc.org/initiatives/mr5/preliminary_recommendations*.

Mark Schwartz, associate professor of medicine at the New York University School of Medicine, elaborated on the importance of evolution in the preparation of future physicians. Medicine is based on biology, and biology is based on evolution, he said, but very few physicians have had the chance to get beyond the basics of evolutionary principles. Furthermore, medical educators and researchers rarely tap into the elegance and power of evolutionary thinking. Undergraduates now have more opportunities than in the past to learn about the interface of evolution, health, and disease, but most premed students have scant room for electives in their schedules. Medical schools have few prerequisites for admission reflecting evolutionary thinking. No North American medical schools require or develop these competencies. As Schwartz said, quoting evolutionary biologist Randolph Nesse, "We are practicing and teaching medicine with only half of biology."

Medical students are intrigued by big questions, said Schwartz. Why do we age? Why do so many of us wear glasses? Why is there a menopause? Why must we sleep? Why do we still have an appendix? Why are autoimmune diseases becoming more common?

Most people in the medical community hear these as proximate, mechanistic "what" questions—how does the body work? As a result, they are drawn to pathophysiologic, mechanistic, or epidemiologic explanations. These explanations are of course important and largely shape the practice of medicine, but they tell only part of the story. "To fully understand the biology of health and disease, one must go beyond these 'what' questions to the evolutionary questions," said Schwartz. Evolutionary questions ask "why." Their answers are framed in terms of selective pressures, phylogenetics, developmental tradeoffs, ecological constraints, and so on.

Evolution provides learners with a conceptual framework, said Schwartz. It integrates basic and clinical sciences, making medical education and practice more coherent. Infusing this integrative science into medical education can foster new questions and insights that provide a sense of discovery about the human condition. "Learners find evolutionary science endlessly intriguing and are quite eager to learn more, but they are very disappointed with the lack of educational opportunities."

Medical schools are complex systems that are very slow to change. Quoting University of Missouri physician Jack Colwill, Schwartz said that medicine educates tomorrow's physicians in today's system while maintaining yesterday's beliefs. Curricular time is of course precious at all levels of education, and many valuable fields are vying for that time to educate premedical and medical students. But teaching and learning about evolution provides a conceptual scaffold on which facts can be organized, not simply a new set of facts. Evolutionary science "provides the bridges and tunnels that students need to connect and navigate what otherwise can sometimes seem like an archipelago of various sciences," said Schwartz.

At the time of the workshop, the National Evolutionary Synthesis Center had just formed a new working group to lay the groundwork for ongoing endeavors to provide testable curriculum models and pathways for infusing evolutionary thinking into premedical and medical education. The working group will refine core competencies across the continuum of premedical to medical training with a focus on teachable moments, since there is not enough room in the medical school curriculum for a whole new course. It will seek to infuse evolutionary thinking into the basic science and clinical education of trainees, and model curricula and learning experiences will be open for all to use. "Of course, these efforts by themselves will not be sufficient, but hopefully this will produce an intellectual platform from which educational interventions, including randomized control studies, can test the efficacy of these interventions."[3]

PROFESSIONAL SOCIETIES

The National Association of Biology Teachers

The National Association of Biology Teachers (NABT) has been changing in recent years, said its executive director, Jaclyn Reeves-Pepin. With a membership of 4,200 people, NABT traditionally has been primarily an organization of high school teachers. However, NABT now consists of about half high school teachers and half two-year and four-year college professors. As a result, the organization is serving a population—higher education faculty—not traditionally associated with the organization. However, it still does not serve most middle school and elementary school teachers, which is where teaching about evolution needs to begin.

Several years ago, NABT did an extensive survey of its members in which it asked the question, "What are the top five topics that NABT members are most interested learning about?" On a list of possible answers, genetics was first, appearing on well more than 50 percent of the lists. Evolution was second, at about 50 percent, followed by environmental science, molecular biology, and human anatomy and physiology.

The second part of that question was, "Which topics do you teach?" with exactly the same options listed as possible answers. Evolution was being taught by fewer than 30 percent of NABT members. "One hundred

[3] About two months after this convocation, the journal *Evolution: Education and Outreach* published a special issue devoted to teaching and learning about evolutionary medicine. That issue can be accessed at *http://www.springerlink.com/content/1936-6426/4/4*.

percent of NABT members should be saying that they are teaching evolution, but they didn't," said Reeves-Pepin.

Since then, NABT has stepped up its work on teaching evolution. It has partnered with multiple organizations, including NASA, BioQUEST, the Smithsonian Institution, and the Howard Hughes Medical Institute on evolution-related initiatives. It has organized talks across the country for parents, school boards, and the public by nationally recognized speakers on how to address controversy and science denialism in the classroom. It also has tried to make its members ambassadors for the profession. "One strong teacher in a district who knows how to address the teaching of evolution and the teaching of biology can become a teacher or leader in that district and can create change at a very local level. We do not want to underestimate the impacts those teachers can have."

In October 2011, NABT released a position statement on teaching evolution.[4] It said:

> Just as nothing makes sense except in the light of evolution, nothing in biology education makes sense without reference to and through coverage of the principles and mechanisms provided by the science of evolution. Therefore, teaching biology in an effective, detailed, and scientifically and pedagogically honest manner requires that evolution be a major theme throughout the life science curriculum both in classroom discussions and in laboratory investigations. . . . Biology educators at all levels must work to encourage the development of and support for standards, curricula, textbooks, and other instructional frameworks that prominently include evolution and its mechanisms and that refrain from confusing non-scientific with scientific explanations in science instruction.

If high school and college teachers do not include evolution as an integral component of their courses, they are doing a disservice not only to their fellow professionals but also to themselves and their students, said Reeves-Pepin. Furthermore, they harm the entire society, she added, because their students will be future voters and future parents.

The American Institute of Biological Sciences

Professional societies can play a major role in encouraging the teaching of evolution across the biology curriculum, said James Collins, Virginia M. Ullman Professor of Natural History at Arizona State University, former assistant director for the Biological Sciences Directorate at National Science Foundation, currently president of the American Insti-

[4] The statement, along with a variety of other resources, is available at *http://www.nabt.org/websites/institution/index.php?p=110.*

tute of Biological Sciences (AIBS), and a member of the convocation's organizing committee. AIBS, which is an umbrella organization with about 160 member societies and 225,000 affiliated scientists (through both individual memberships in AIBS and through their membership societies), has maintained a strong and persistent presence in advancing evolutionary biology. The journal *BioScience* is a major organ for disseminating information to the community, and especially to educators. The website ActionBioscience.org, which seeks to "bring biology to informed decision making," is another way in which AIBS communicates with a larger set of communities. In addition, AIBS has a strong policy presence in Washington, DC, and is involved in alliances on evolution-related issues. "It will be increasingly important for scientific societies to be able to step up and take policy positions, educate people, and participate in the policy arena," said Collins.

The Federation of American Societies for Experimental Biology

The Federation of American Societies for Experimental Biology (FASEB) is also an umbrella organization that represents 24 societies with about 100,000 members collectively, which are involved largely in medical research, said its president Joseph LaManna, professor of physiology and biophysics, neurology, neurosciences, and pathology at Case Western Reserve University. Many of FASEB's member societies have a specific and deep interest in evolution, and all of them work in sciences that involve evolutionary thinking. The organization's website (*http://www.faseb.org*) contains a variety of resources, including many that are related to evolution. FASEB also offers resources for evolution education that include background information and tips and tools for communicating about evolution. It runs conferences, publishes a journal, and has a society management service. It even produces buttons to hand out at scientific meetings with slogans such as "Teach Evolution" and "Take a Stand for Science."

FASEB focuses on science policy, so it remains vigilant for policy issues that affect the teaching of evolution. It has a Capitol Hill office and organizes regular visits with Members of Congress and their staff members. It particularly emphasizes the importance of maintaining a good stream of well-trained researchers to support the nation's research and development base, which requires that students learn evolutionary concepts and content.

In 2005, FASEB's Board of Directors adopted a statement on evolution that reads:

- FASEB considers evolution a critical topic in science education and strongly supports the teaching of evolution.
- FASEB opposes mandating the introduction of creationism, intelligent design, and other non-scientific concepts into the curricula of science.
- FASEB opposes introducing false controversies regarding evolution or other accepted scientific theories into the curricula of science.
- FASEB calls upon the scientific community and American citizens to defend science education by opposing initiatives to teach intelligent design, creationism, and other non-scientific beliefs in science class.

A useful step forward, said LaManna, would be for the scientific societies to do a "meta-review" of available educational resources. Can these resources be aligned and strengthened, so that people are less confused by the wealth and variable characteristics of the available resources?

In the same vein, consolidating efforts may produce more effective initiatives. For example, organizations could partner on policy objectives by bringing people to advocate on Capitol Hill who are not normally represented in discussions on evolution education.

The American Society for Microbiology

Finally, Amy Chang, education director for the American Society for Microbiology (ASM), spoke about four areas in which professional societies have a role: advocacy, guidelines and models, professional development, and information dissemination. ASM has a diverse membership of about 40,000 people, with about 60 percent from colleges and universities and the other 40 percent from companies, federal and state governments, public health laboratories, diagnostic laboratories, and other organizations.

The ASM has a Statement on the Scientific Basis for Evolution[5] that "sets a vision for where we need to be," said Chang. A particular value of such statements is that developing them requires discussion within an organization, which helps unify its intent and initiatives. The National Center for Science Education has a website that links to all similar statements,[6] and Chang urged societies that do not yet have such a statement to generate one.

In the area of guidelines and models, ASM has been developing guidelines for a recommended core curriculum. The introductory course

---

[5] See *http://www.asm.org/images/Education/asm%20evolution%20statement_6_06b.pdf.*

[6] These statements are available at *http://ncse.com/media/voices/science.*

in microbiology, which is the foundation of the curriculum, has six organizing themes, the first of which is evolution. Introductory microbiology is in part a service course for the nursing and allied health professions, and the need for curriculum guidelines differs significantly between majors. The structure of the courses for science majors and the allied health professions is based on employment and passing licensure examinations that are content dense.

ASM also provides professional development in education for its members. Chang estimated that 80 percent of the society's members have opportunities to explain science to people in their communities, including parents, youth, children, churchgoers, health professionals, science fair or county fair participants, and many others. "They all have an opportunity to bring science to the citizens," said Chang. ASM has been developing "training materials and leadership to empower the members to do their jobs in explaining the theories of evolution or evolutionary science and bringing it to everyday life."

Finally, ASM is involved in information dissemination through a variety of publications and other documents.

RESOURCES FOR TEACHING EVOLUTION ACROSS THE CURRICULUM

The Understanding Evolution Website[7]

The Understanding Evolution website, which was launched in January 2004 with support from the National Science Foundation, was designed to give K-12 teachers the content knowledge and resources to teach evolution with confidence. The developers of the website quickly realized that its potential audience was much larger than teachers, said Judy Scotchmoor, assistant director of the University of California Museum of Paleontology. With additional funding from the Howard Hughes Medical Institute, a new version of the site launched a year and a half later. The teacher site was still available, but university instructors, students, and the general public also were included as intended audiences.

In January 2011, a new version of the website launched in partnership with AIBS and NESCent. The site contains teaching materials, a resource library, an in-depth online course on the science of evolution, and stories on how evolution factors into current news stories. The site has a large international audience and has been translated into multiple languages. As the site says, it "is here to help you understand what evolution is, how

[7] Available at *http://evolution.berkeley.edu/*.

it works, how it factors into your life, how research in evolutionary biology is performed, and how ideas in this area have changed over time."

Several factors have contributed to changes in the website over the past eight years, said Scotchmoor. One was the discovery that about 30 percent of the site's audience taught at the undergraduate level, even though the target audience for the first version of the site was solely K-12 teachers. Another was input from a high-level advisory group that met in 2008. Scotchmoor recalled a particular comment from group member Rodger Bybee: "Overnight, you're not going to get everybody to suddenly start integrating evolution into the teaching of biology wherever they are. But if we can encourage them to take baby steps and that first step is comfortable, they'll take another step."

Another critical factor was the successful submission of a curriculum development grant to the National Science Foundation in 2009. The goals of the grant were to:

- Encourage college biology instructors to integrate evolutionary concepts—especially the applications and relevance of evolution—throughout their biology teaching.
- Encourage college biology instructors to spend more class time on evolution-related concepts and emphasize the currency of evolution research in their instruction.
- Encourage college biology instructors to use pedagogical techniques supported by education research in their evolution instruction.
- Impact college students.

With funding from that grant, the Understanding Evolution team put together the Understanding Evolution Teacher Advisory Board, which has helped improve the site in many ways. Board members showed how to make the navigation and access better. They worked on how to overcome some of the reasons teachers give for not teaching evolution across the curriculum, which led to development of the "Evolution 101" course that the website provides. The board made sure that additional resources for any given topic in evolutionary science were easily and readily accessible. These resources include examples, case studies, teaching recommendations, and research profiles of evolutionary biologists and their research.

The teaching materials have changed dramatically over the lifetime of the site. The site now has "teachers' lounges" at the K-2, 3-5, 6-8, 9-12, and undergraduate levels. Each lounge opens with four bullets that link to information appropriate for that grade level:

- Focus on the fundamentals
- Identify your learning goals

- Avoid common teaching pitfalls
- Search for lessons

All of these grade levels are important, said Scotchmoor. A second-grade teacher may say, "I don't teach evolution." But even in second grade, students can observe that not all kittens look the same and that they inherit certain characteristics from their parents.

By condensing many different syllabi from introductory biology courses, the developers of the Understanding Evolution site also have put together an "interactive syllabus" that links evolution to topics throughout the biology curriculum. For example, when teaching the Krebs cycle, the site provides five-minute slide sets that connect the topic to evolution. The interactive syllabus also provides teaching tips, learning goals, and modifiable teaching scripts for different topics.

The group has been investigating slide sets that promote active learning to incorporate into the syllabus. A journal toolkit enables teachers and students to access the primary literature. Finally, an Evo Gallery provides students with the opportunity to use a medium they choose to talk about evolutionary concept. Students peer review each other's creations, and their contributions are archived on the website so that the selection continues to grow.

The take-home messages, said Scotchmoor, are to make information easy and accessible, provide appropriate packaging and guidelines for use, create modifiable formats for different teaching styles, provide resources that also target other content and skills that need to be taught, provide assessment and diagnostics whenever possible, engage students actively, provide resources that are relevant to students, and provide professional development for teachers.

Through their experiences with the Understanding Evolution site, the developers realized that a segment of the population is confused about evolution because they are confused about science. The result was a sister site called Understanding Science, which was developed through a multidisciplinary collaboration of scientists and educators.[8] The website is based on three major principles. First, the processes of science have to be explicitly and independently emphasized. Second, throughout instruction, students should be encouraged to examine, test, and revise their ideas about what science is (and is not) and how it works. Third, key concepts about the nature and processes of science should be revisited in multiple contexts throughout the school year, allowing students to see how they apply in real-world situations. "Introducing the process of sci-

[8] Available at *http://understandingscience.org*.

ence in pages three through five of any textbook and then forgetting about it is not going to help," Scotchmoor said.

During the discussion period following this panel discussion, LaManna suggested that the ideas in Understanding Evolution be linked to Wikipedia, particularly in areas where evolution has an impact on daily life, an idea Scotchmoor labeled brilliant. She also responded to a question by saying that members of the website's Teacher Advisory Board consider their work to be professional development, and many of them do it essentially as volunteers.

The National Evolutionary Synthesis Center

The National Evolutionary Synthesis Center (NESCent) responds to the interests and the goals of the evolutionary biology community, said Kristin Jenkins, who works on education and outreach with the organization and who served as a member of the convocation's organizing committee. With funding from the National Science Foundation, NESCent has a broad mandate and has developed educational materials for diverse audiences. Most of its products are focused on high school and college levels, because that is where evolution is commonly taught. In addition, proposals from the community lead to the formation of working groups that pursue new initiatives. Some working groups develop particular materials, like assessments. Others focus on specific problem areas such as tree thinking. And some work on big picture ideas, like the Teaching Evolution Across the Curriculum Working Group that served as the impetus and catalyst for the development of this convocation.

NESCent has been thinking about how the evolutionary approaches taught in biology can be applied in other areas. "Many of the students in introductory biology are not necessarily going to be biology majors," said Jenkins, "but having them pick up that way of thinking and being able to use that in their future careers, as well as being aware of how biology works, is very important to NESCent."

NESCent also provides professional development so that teachers are knowledgeable and confident in teaching evolution. It works with NABT and AIBS to offer a symposium on evolution every year at the NABT annual meeting.[9] And it partners with groups such as Understanding Evolution and BioQUEST to develop specific materials so that if faculty members want to try something novel, they have the support and the resources to do so.

Finally, NESCent works directly with students through such groups as the Society for the Advancement of Chicanos and Native Americans in

[9] Links to all of these symposia are available at *http://www.aibs.org/events/special-symposia/*.

Science (SACNAS)[10]—for example, by having scientists talk with students to keep them engaged with science. "There are a lot of opportunities to provide scaffolding and support for the people who are in the classroom," Jenkins said.

Textbooks

One of the objections from William Buckingham, who was the head of the school board's curriculum committee in Dover, Pennsylvania, about the textbook *Biology* (Miller and Levine, 2004) is that the book is "laced with Darwinism" from beginning to end. He was objecting to the fact, said Joseph Levine, one of the textbook's authors, that, similar to the school superintendent in Kentucky who ordered two pages of a textbook glued together (see Chapter 1), a few pages could not be glued together to eliminate mention of evolution. Rather, Levine emphasized, evolution is intentionally integrated throughout the book.

The treatment of evolution has increased in successive editions of the book, said Levine, who has been publishing the book with his coauthor Kenneth Miller for more than two decades with Prentice-Hall (now Pearson Education). The first edition had 63 pages in the evolution unit and 17 references in the index to evolution, partly because the publisher was "afraid of putting too many in," said Levine. The current edition of the textbook has 123 pages in the unit on evolution and 45 index entries under evolution. Furthermore, many of the chapters in the book have an obvious evolutionary perspective, while others have what Levine referred to as "stealth inclusions" in such areas as genetics and molecular biology. The most recent edition has updated coverage of phylogenetics and a discussion of cladistics. It also observes that the protista are not a kingdom. "Biologists haven't thought that for about 30, 35 years now. The problem is that most state standards still refer to the kingdom protista, and teachers are obliged to teach about that."

The book emphasizes concepts rather than facts. Each section of the book starts with key questions that are conceptually based. New vocabulary comes later and is designed to serve the concepts. "We don't start off the paragraph with a foreign language word that knocks the kids off balance. We start out by discussing what we're talking about and then put a name to it."

For any educational program to make a difference in K-12 education on a national scale, it has to succeed in the marketplace, Levine emphasized. The very best materials will have limited value if only the top 10 percent of teachers and the top 10 percent of students get to see them.

---

[10] Additional information is available at *http://sacnas.org/*.

That top 10 percent may include the students headed to medical school or research. But the other 90 percent, who are not honors or AP students, become the large majority of the general public. Thus, Levine and Miller have had to continually upgrade and improve the presentation of evolutionary thinking while working to ensure that the book is not banned in the marketplace.

Levine described some lessons drawn from his experience as a textbook author. First, implementing materials in K-12 education requires patience as teachers assimilate new material, a willingness to negotiate, and a major investment of time and energy in conceptualizing materials and making sure that they are used and have an effect. The book has many different kinds of ancillary materials, such as English language learner support and differentiated instruction. These materials "require an enormous up-front investment, and this is a commercial operation so they have to recoup that." Also, national and state standards are much more important than most people realize. "The most powerful selection pressure in the marketplace is state standards and the assessments on which teachers and their students are evaluated." In fact, evaluation rubrics in some states penalize books not only for not including state standards but also for including material that is not in the standards. The result is "an enormous pressure on publishers and on creators of materials in terms of conforming to the standards."

The most interesting, up-to-date, relevant, and important evolutionary subjects and cross-connections across the biological disciplines are not in most current curricula and state assessments. The new National Research Council framework (2011) "looks fantastic," according to Levine. But to benefit from the work that has gone into the framework, standards need to be developed that states will adopt, and groups of teams then will need to work on a state-by-state basis to shepherd those standards into curricula. "Others have done very good standards, [but] by the time they got down to the teachers, they were lists of vocabulary words."

In addition, many of the people teaching biology today are poorly suited to teach about science through inquiry because they are reluctant to lead their students into areas where they may not know all the answers. "They're not Mr. or Ms. know-it-all anymore. We have to think creatively about lots of new kinds of communication and professional development."

Finally, Levine mentioned that the website of a PBS series on evolution for which he served as science editor about a decade ago still exists and is regularly updated.[11] "It's a fantastic resource that is involved in communicating to a more general audience than most of the people in

[11] See *http://www.pbs.org/wgbh/nova/evolution.*

this room are accustomed to doing." He also is involved in an effort to create new kinds of inquiry-based professional development for teachers.

Evolution Laboratories

If a course in evolution does not have a laboratory component, students confuse the subject with philosophy and religion, because other biology courses have labs, said John Jungck, Mead Chair of the Sciences at Beloit College and the originator of BioQUEST.[12] Jungck has been involved in setting up a variety of evolution laboratories using such tools as phylogenetic trees, bioinformatics, multivariate statistics, exploration of real biological databases, or simply biological variation. All of these options can include field work. "If you want [students] to love biodiversity, get them out in it." Even students at a very young age can be engaged in biological diversity, and older students can contribute to original research.

As an example, Jungck described the BIRDD approach to evolution labs, where BIRDD stands for Beagle Investigations Return with Darwin Data. Undergraduate students work with original data from the finches Darwin studied in the Galapagos Islands on such characteristics as wing length, upper beak length, bird songs, and georeference maps. They also use modern data such as phylogenetic trees, protein sequences, and nucleic acid sequences. In one student project, three students used multivariate statistics to build a three-dimensional plot based on just three measurements of the physical characteristics of the 13 Galapagos finch species. They then examined character displacement with populations that overlap and are geographically separated. "It was beautiful," said Jungck. "Students are testing evolutionary theory with data, and they have the pride of ownership of their investigation and their products."

In another experiment, students investigate HIV data from 600 patient visits in Baltimore to study the evolution of protein structure and function. Jungck also briefly described an investigation involving measurements of sea shells. Over the course of evolution, sea shells have taken some shapes but not others, which is an observation that students can make for themselves. It is then possible to engage them in discussions of questions such as why some shapes are absent and why some forms appear in the geologic past but are no longer observed in extant species.

"We can engage students with real-world data and real-world questions," said Jungck. "They are investigators. They're coming to learn science and do science."

---

[12] Additional information is available at *http://bioquest.org*.

REFERENCES

Brewer, C., and Smith, D., Eds. 2011. *Vision and Change in Undergraduate Biology Education: A Call to Action*. Washington, DC: American Association for the Advancement of Science.

Miller, K. R., and Levine, J. S. 2004. *Biology*. Upper Saddle River, NJ: Prentice-Hall.

National Research Council. 2002. *Learning and Understanding: Improving Advanced Study of Mathematics and Science in U.S. High Schools*. J. P. Gollub, M. Bertenthal, J. Labov, P. C. Curtis, Eds. Washington, DC: National Academy Press.

National Research Council. 2011. *A Framework for K-12 Science Education: Practices, Crosscutting Concepts, and Core Ideas*. Washington, DC: The National Academies Press.

7

Next Steps

During the final breakout session, convocation participants met in four groups organized by areas of expertise—faculty, funders, education researchers and professional developers, and representatives from professional societies—and discussed the actions that need to be taken to advance and implement the idea of teaching evolution across the curriculum. They then presented and discussed these actions during the final plenary session of the convocation. The proposed actions listed below consist of suggestions made by individuals at the convocation. They should not be seen as consensus conclusions of the meeting or as positions that are officially endorsed by the National Research Council or the National Academy of Sciences. They reflect the conversations that occurred during the convocation and point in some particularly interesting and promising directions.

PROFESSIONAL DEVELOPMENT

- Instructors at all educational levels need continual professional development to be able to teach evolution across the biology curriculum.
- Local, regional, or national academies on teaching evolution across the curriculum could introduce educators to the idea, inform them about available resources, and provide them with support.
- A campaign on intentionally teaching about evolution and the nature of science every day could bring the idea to all educators and education administrators.

RESOURCES

- A clearinghouse of resources related to evolution education could disseminate existing materials, foster the development of new materials, and catalyze the creation of networks of biology educators. In particular, a compilation of best practices would be a valuable resource not only for biology educators but also for instructors in other subjects who want to incorporate evolution into their teaching.
- Existing resources such as the Understanding Evolution website could be expanded and promoted. In particular, these resources could present examples provided by disciplinary and professional societies across the life sciences and reviewed for pedagogical effectiveness and potential impact by various teacher organizations such as NABT, the Understanding Evolution websites' Teacher Advisory Board, or the National Academies' Teacher Advisory Council.[1]
- A taskforce supported by the National Academy of Sciences or other scientific organizations could develop materials on evolution and the nature of science both for educators and the general public.
- A searchable database of curated education research papers could make what is known about the teaching of evolution available to all instructors.
- Compilation and publication of known effective techniques (the "hooks" to which Robert Pennock referred during his presentation and discussed in other parts of this report) in formats that are readily available and easily accessible to teachers, including what might be learned from collaborating with international science educators about best practices in other parts of the world.
- Establishment during the coming year of an organizing body that would spearhead the development and operation of a clearinghouse of resources, research practices, metrics for measuring both student learning and the efficacy of programs, strategies for more effective teaching of evolution, and coordination with media for efforts such as the "Everyday Evolution" initiative suggested by Uno during his presentation.

[1] Additional information about the National Academies Teacher Advisory Council is available at *http://www7.nationalacademies.org/tac*.

RESEARCH

- Additional education research could investigate how students at all ages learn about evolution and the best ways of conveying information about the subject.
- The coordination of research throughout the country could enable educators and education researchers to work together to generate new knowledge on evolution education.

PUBLIC OUTREACH

- Articulating classroom-based education efforts with informal learning environments could reinforce and extend the teaching of evolution across the curriculum.
- Educators can be ambassadors for both evolution and the nature of science by volunteering for committees, talking with colleagues, speaking in public events, publishing articles, and engaging in other outreach efforts.
- Strategically planned and financed dissemination of the ideas of "evolution every day" or "everyday evolution" to a variety of audiences could build awareness of the centrality of evolution in the modern understanding of life.
- A watchdog group could rate politicians for scientific accuracy just as other groups rank politicians on other issues.

SUPPORT

- Supplemental awards from the National Science Foundation and other public and private sources of funding can provide support for educational and research activities involving evolution education, and especially for students who do not take Advanced Placement Biology and those who are not planning to major in science.
- NESCent can support the development of courses and curriculum for evolution education at different levels, from elementary school to college and from the local to national scales.
- Deleting some material from current biology curricula may be necessary to make room for an increased focus on evolution.
- Rewarding college faculty for effective teaching about evolution and the nature of science could create incentives to develop new materials and teach students well.

PROFESSIONAL SOCIETIES

- Continued strong support from the National Academy of Sciences and other scientific organizations can provide encouragement for teaching evolution in high schools and tying evolution to national standards.
- Professional societies could specifically target and recruit the participation of teachers from high schools and community colleges to their annual meetings to explore how evolution is central to their disciplines and how evolution can be better integrated into appropriate sections of biology courses in those disciplines.
- Professional societies could support presentations at annual meetings of educators on topics related to evolution education.

Robert Pennock from Michigan State University expressed an appropriate closing comment: "We are at a cusp where the communities are coming together to teach evolution across the curriculum. We've been wanting to do this for a long time. I hope that, a little while from now, we can look back on this [convocation] as something to celebrate."

Appendixes

A

Convocation Agenda

Thinking Evolutionarily:
Evolution Education Across the Life Sciences
Organized by:
Board on Life Sciences, National Research Council
National Academy of Sciences
Co-hosted by Carnegie Institution for Science
1530 P St., NW, Washington, DC
October 25-26, 2011

We sincerely thank the National Academy of Sciences, the Burroughs-Wellcome Fund, the Christian A. Johnson Endeavor Foundation, the Carnegie Institution for Science, and the National Science Foundation through a Research Coordination Network/Undergraduate Biology Education grant to the University of Oklahoma for their generous support of this convocation.

DAY 1: EXPLORING THE OPPORTUNITIES AND SETTING THE STAGE FOR FUTURE ACTION

Tuesday, October 25

11:15 AM	**Registration**, First Floor Foyer **Lunch** Available beginning at 11:30, Rotunda (2nd floor)
12:00 PM **Auditorium**	**Welcome and Introductions** - Jay Labov, National Research Council and National Academy of Sciences - Maxine Singer, President Emerita, Carnegie Institution for Science

- Susan Kassouf, Program Officer, Christian A. Johnson Endeavor Foundation (sponsor)

- Cynthia Beall, Chair of the Convocation's Organizing Committee and Moderator

12:15 PM
Auditorium **The Case for Thinking Evolutionarily Across the Life Sciences**

- Introductory Undergraduate Biology Courses: Gordon Uno, University of Oklahoma and PI for the NSF's Research Coordination Network for Undergraduate Biology Education (special advisor to the Committee and PI of the NSF/RCN-UBE grant that is sponsoring this convocation).

- Judy Scotchmoor, Museum of Paleontology, University of California, Berkeley

Questions and Discussion

1:15 PM
Auditorium **Can This Approach Improve Student Learning of Evolution? The Evidence Base**

- Ross Nehm, Ohio State University

Questions and Discussion

2:00 PM
Rotunda **Break and opportunity for further networking**

2:15 PM
Auditorium **Expanding Curricular Opportunities to Introduce Evolutionary Thinking Across the Grade Spans—Brief Presentations and Panel Discussion**

- Spencer Benson, University of Maryland: *The Role of Evolution in the Restructured Advanced Placement Biology Course*

- Celeste Carter, National Science Foundation: *Vision and Change in Undergraduate Education*

- William Galey, Howard Hughes Medical Institute: *Scientific Foundations for Future Physicians*

- Kristin Jenkins, National Evolutionary Synthesis Center and Member of the Organizing Committee: *NESCENT Programs Promoting Evolutionary Thinking*

- Mark Schwartz, New York University: *Evolutionary Medicine in Biology and Pre-Med Courses*

Discussion and Questions

3:15 PM
Auditorium **Who Are the Audiences We Are Trying to Reach with this Initiative?**

- Cynthia Beall, moderator

3:45 PM
Auditorium **How Can Evolutionary Thinking Help Address the Controversies Surrounding the Teaching of Evolution?: A Faculty Forum**

- Betty Carvellas, National Academies Teacher Advisory Council

- David Hillis, University of Texas, Austin

- Paul Strode, Fairview High School (Boulder, CO)

- Marlene Zuk, University of California, Riverside

4:30 PM
Breakout
Rooms **First Breakout Sessions: Exploring the Issues In Greater Depth**

The colored dot on your name badge indicates the breakout session to which you have been assigned. Each of these breakout sessions will contain a mix of people with different kinds of expertise. The goal of each session will be to explore in depth one of the issues raised in earlier sessions and report back ideas for next steps to all participants. Each group will be facilitated by a member of the organizing committee. Each group will appoint one person to present an overview of the group's ideas and suggestions at the end of the morning. Each group will decide when to call a break.

Group 1 (yellow dot, Ballroom): What constitutes evolutionary thinking? What approaches are needed to educate faculty and departments about the value of evolutionary thinking in their own courses and programs?
Facilitated by Nancy Moran, Yale University

Group 2 (blue dot, Board Room): What additional evidence is needed to convince biologists of the value of evolutionary thinking? How can that evidence best be gathered through an organized program of research? Who should undertake and sponsor such research?
Facilitated by Ida Chow, Society for Developmental Biology, and Paul Beardsley, California Polytechnic University

Group 3 (green dot, Mayor Room): How can evolutionary thinking become more firmly connected with other emerging efforts to improve life sciences education? In what ways should these efforts be influenced by different target audiences?
Facilitated by Gordon Uno, University of Oklahoma, and Kristin Jenkins, National Evolutionary Synthesis Center

5:30 PM
Auditorium **Reports from Breakout Groups (10 minutes each plus discussion)**

6:15 PM
Auditorium **Closing Remarks, Announcements, and Charge for Day 2**

- Cynthia Beall

6:30 PM
Auditorium **Adjourn for the Day**

Evening **Dinner on Your Own for Participants (see accompanying list of suggested restaurants in the Dupont Circle area)**

DAY 2:
PLANNING FOR FUTURE ACTIONS TO INFUSE EVOLUTIONARY THINKING ACROSS LIFE SCIENCES EDUCATION

Wednesday, October 26

7:30 AM **Breakfast**
Available in the Rotunda

8:00 AM
Auditorium **Synthesis, Reflections on Day 1 and on Moving Forward**

- Robert Pennock, BEACON Center for the Study of Evolution in Action, Michigan State University

8:45 AM
Auditorium **Reactions and Further Discussion**

- Panel of Committee Members- Open microphone for participants

9:15 AM
Auditorium **Expanding Resources for Teaching Evolutionary Thinking**

- Paul Beardsley (member of the organizing committee), California Polytechnic Institute

- Joseph Levine, Pearson Education and Co-Author (with Kenneth Miller) of *Biology*

- Judy Scotchmoor, Director, Understanding Evolution and Understanding Science Websites, Museum of Paleontology, University of California, Berkeley

- John Jungck (Beloit College, *BioQuest*)

Discussion and Questions

10:00 AM
Auditorium **Moderated Panel Discussion: Next Steps: Potential Roles of Key Players**

- James Collins (Arizona State University and member of the organizing committee), President of American Institute of Biological Sciences

- Jaclyn Reeves-Pepin, Executive Director, National Association of Biology Teachers

- Joseph LaManna, President, Federation of American Societies for Experimental Biology

- Amy Chang, Director of Education Programs, American Society for Microbiology

10:30 AM
Rotunda **Break and opportunity for further networking**

11:00 AM
Auditorium **Moving Evolution Education Forward: Why Evolution and Evolutionary Thinking Are Integral Components of *Molecular Biology of the Cell***

- Bruce Alberts, University of California San Francisco, Editor-in-Chief, *Science*

Questions and Discussion

12:00 PM Breakout Rooms – Lunch is available in the Rotunda **Second Breakout Sessions: Moving from Vision to Action (Working Lunch)**
These sessions are designed to have people with similar interests and expertise meet with each other to craft action items that can be carried forward. Your group should: develop up to three action items that can be undertaken by colleagues in the sector your group represents in the next six months; discuss how your action items might connect with at least one of the other

sectors represented in breakout groups; and discuss how the National Research Council and National Academy of Sciences might assist your efforts.

Group 1 (Board Room): Faculty who teach courses in biology and evolution.
Facilitated by Irene Eckstrand, National Institutes of Health, and Nancy Moran, Yale University

Group 2 (Mayor Room): Funders of programs in life sciences education.
Facilitated by James Collins, Arizona State University, and Kristin Jenkins, National Evolutionary Synthesis Center

Group 3 (Ballroom): Representatives from Professional Societies.
Facilitated by Ida Chow, Society for Developmental Biology

Group 4 (Auditorium): Curriculum Developers and Education Researchers.
Facilitated by Gordon Uno, University of Oklahoma, and Paul Beardsley, California Polytechnic Institute.

1:30 PM **Reports from Breakout Groups (10 minutes each plus discussion)**

2:30 PM **Closing Thoughts and Reflections**

- Members of the Organizing Committee, Other Participants

3:00 PM **Adjourn**

B

Brief Biographies of Committee Members and Staff

Chair
Cynthia M. Beall (Member, National Academy of Sciences)
S. Idell Pyle Professor of Anthropology, Case Western Reserve University

Cynthia Beall's primary research focuses on how populations with different microevolutionary histories adapt to the lifelong environmental stress of high-altitude hypoxia. She conducts her research with populations on the Andean plateau of South America, the Tibetan Plateau of Central Asia, and the Simien Plateau of East Africa. This work has revealed two different patterns of adaptation to hypoxia rather than the single universal human response envisioned by classical environmental physiologists. She also investigates how the influence of the sociocultural environment can both create and buffer stress and can have beneficial and detrimental effects on human biology.

She received her bachelor's degree in biology from the University of Pennsylvania and her M.S. and Ph.D. degrees in anthropology from Pennsylvania State University.

Dr. Beall also has a long-term interest in evolutionary medicine and is working with colleagues on her campus and with the National Evolutionary Synthesis Center to develop resources for teaching this topic to undergraduates.

She has a long record of service to the NAS, including the NAS Council, and the NRC. Currently Dr. Beall is a member of the Division Committee for the Division on Behavioral and Social Sciences and Education.

Paul Beardsley
Center for Excellence in Mathematics and Science Teaching (CEMaST)
California State Polytechnic University, Pomona

Paul Beardsley recently joined the faculty at California State Polytechnic University, Pomona. Previously, he was a science educator at the Biological Sciences Curriculum Study (BSCS) where he was the principal investigator (PI) for the NIH-sponsored Evolution and Medicine high school project and Co-PI for the Rare Diseases and Scientific Inquiry middle school project. He worked on high school and middle school comprehensive curricula, including BSCS Science: An Inquiry Approach, BSCS Biology: A Human Approach, and Agile Mind Biology. He helped design the on-line professional development program for teachers of multidisciplinary science called Across the Sciences (a joint project with Oregon Public Broadcasting) and professional development with a range of teachers, including developing a leadership institute with Seattle Public School science teachers. He is also currently conducting research in evolution education.

In prior faculty positions at Idaho State University and Colorado College, Dr. Beardsley helped direct a doctoral-level program in biological education and participated in a NSF GK-12 program matching graduate students in science with K-12 teachers. He taught graduate courses in biology education and educational research, trained pre-service teachers, and also taught graduate and undergraduate courses in evolution, ecology, botany, and cell biology. Graduate students in his lab carried out research in plant evolution and biology education.

Dr. Beardsley earned a Ph.D. degree at the University of Washington in plant evolution. His research focuses on plant molecular systematics, the genetics of plant speciation, and the genetics of rare plants. He also earned a secondary science teaching license and taught at both the middle and high school levels. His research interests in science education include student learning in evolution and scientist-teacher partnerships.

Ida Chow
Executive Officer
Society for Developmental Biology

Ida Chow is the executive officer of the Society for Developmental Biology (SDB). She manages the society and participates in many of its activities, including educational activities such as education symposia and workshops at the annual and regional meetings, Boot Camp for New Faculty, a teaching digital library, science education outreach at all levels, and career issues. She organized SDB's "perfect partner" participation at the First USA Science and Engineering Festival held in October 2010, with a teacher workshop, a speaker at the Nifty Fifty series, a Nobel laureates lecture and pre-lecture meeting with local students and their parents, and a booth with viewing of live frog and zebrafish embryos and other hands-on activities at the Festival Expo.

Dr. Chow was Co-PI of three NSF Pan American Advanced Studies Institute (PASI) program grants to conduct short courses for graduate students and postdoctoral fellows from U.S. and Latin American institutions in developmental biology in Brazil (2005), Argentina (2008), and Chile (2010), a collaboration between SDB and the Latin American Society for Developmental Biology. She also was the author of an NSF sub-award for SDB's teaching digital library, LEADER, a partner of the BEN Pathway administered by the American Association for the Advancement of Science. She chairs the Coalition of Scientific Societies composed of more than 30 scientific and professional societies, which focuses on teaching evolution for all; and she coordinated participation of 16 of these societies in a common activity, the Evolution Thought Trail at the Expo of the 2010 Science and Engineering Festival.

Dr. Chow received her bachelor's degree in biomedical sciences from Escola Paulista de Medicina in São Paulo, Brazil, and Master's and Ph.D. degrees from McGill University in Montreal, Canada. She held research and teaching positions at University of California Irvine, University of California Los Angeles, and American University (Washington, DC) before joining SDB. She was elected an American Association for the Advancement of Science Fellow December 2010.

James P. Collins
Virginia M. Ullman Professor of Natural History and the Environment, School of Life Sciences, Arizona State University

James Collins has been a faculty member at Arizona State University (ASU) since 1975. His research group studies host-pathogen biology and its relationship to the decline of species, at times even to extinction. Dr. Collins' research also focuses on the intellectual and institutional factors that have shaped ecology's development as a science as well as ecological ethics. Dr. Collins was founding director of ASU's Undergraduate Biology Enrichment Program, and served as co-director of ASU's Undergraduate Mentoring in Environmental Biology and Minority Access to Research Careers programs. He has been chairman of the Zoology, then Biology Department at ASU. At the National Science Foundation he was director of the Population Biology and Physiological Ecology Program (1985-1986) and assistant director for biological sciences (2005-2009). Dr. Collins has a Ph.D. from the University of Michigan and a B.S. from Manhattan College. He is an elected fellow of AAAS and the Association for Women in Science and president of the American Institute of Biological Sciences.

Irene Eckstrand
Program Officer, Models of Infectious Disease
National Institute of General Medical Sciences, National Institutes of Health

Irene Eckstrand specializes in evolutionary biology, genetics, and computational biology. As a program director at National Institute of General Medical Sciences, she manages grants that promote research in these areas and directs a program that promotes computational and mathematical research to detect, control, and prevent emerging infectious diseases. The program, called MIDAS (for Models of Infectious Disease Agent Study), was founded in 2004 with the aim of improving the nation's ability to respond to biological threats promptly and effectively. Dr. Eckstrand also manages a new consortium to develop models of the dynamics of the scientific workforce and handles NIGMS research focused on evolutionary biology, including how pathogens and hosts evolve together; speciation; and the evolution of complex biological systems.

From 1999-2004, Dr. Eckstrand directed the Bridges to the Future Program, part of the NIGMS Minority Opportunities in Research Division. The program assists minority students in making the transition to baccalaureate and doctoral programs and prepares them for careers in biomedical research. In the mid 1990s, Dr. Eckstrand directed NIH's Office of Science Education and has worked with professional societies, including the Society for the Study of Evolution and other groups to promote effective biology and mathematics education.

Dr. Eckstrand received a bachelor's degree from Earlham College, a master's degree from Wright State University, and a Ph.D. from the University of Texas at Austin.

Kristin Jenkins
Education and Outreach, National Evolutionary Synthesis Center

Kristin Jenkins works with the National Evolutionary Synthesis Center (NESCent) and the BioQUEST Curriculum Consortium to pursue her interests in biology education. Her experience includes teaching at the high school and college levels, professional development for K-14 faculty, curriculum development, and development of outreach programs. Currently, she is part of the Education and Outreach group at NESCent, where she has participated in various working groups focused on enhancing evolution education including Evolution Across the Curriculum, Tree Thinking in Evolution Education, and Communicating Human Evolution. As a member of the BioQUEST staff, she is part of the Cyberlearning for Community Colleges project and other BioQUEST projects. Dr. Jenkins is an active member of the Society for the Study of Evolution's (SSE's) Education Committee, working with colleagues to provide professional development workshops and symposia to both K-12 teachers and SSE members. She is on the Editorial Board for the journal *Evolution: Education and Outreach*, and is the chair of the Outreach Committee for the University of Wisconsin's J.F. Crow Institute for the Study of Evolution. She earned her B.A. in biochemistry and molecular biology at the University of California, San Diego and her Ph.D. at the University of Arizona.

Nancy A. Moran (Member, National Academy of Sciences)
William H. Fleming Professor of Biology, Yale University

Nancy Moran is a leader in the study of the evolution of symbioses between multicellular organisms and microbes. She uses a variety of genetic, genomic, and biomolecular tools in studying symbioses, focusing on symbioses found in plant-feeding insects. This research is part of her broader interests in the evolution of biological complexity, as found in complex life histories, interactions among species, and in species-diversity of larger biological communities. She works with students and postdoctoral associates to identify the impacts of microbial symbionts on host survival and reproduction, the evolutionary origins of microbial symbionts from free-living bacteria, and genomic changes in evolving symbiont lineages. For her research on symbiosis, she received the 2010 International Prize for Biology from the Japanese Society for the Promotion of Science.

Dr. Moran received her bachelor's degree from the University of Texas and her M.S. and Ph.D. degrees in zoology from the University of Michigan.

Dr. Moran has long taught evolutionary biology to undergraduate students and originated a high school biology research program at Tucson High School. She has a long record of service on NAS committees. Most relevant to her nomination to this committee is her service on the authoring committee for the 2008 NAS/IOM publication, *Science, Evolution, and Creationism.*

SPECIAL CONSULTANT TO THE PROJECT

Dr. Gordon E. Uno joined the Department of Botany and Microbiology at the University of Oklahoma in 1979, was appointed a David Ross Boyd Professor of Botany in 1997, and is currently serving as the Department's chair. Dr. Uno has authored or co-authored several textbooks including: *Principles of Botany; Handbook for Developing Undergraduate Science Courses; Biological Science: An Ecological Approach* (a high school biology text for which he served as editor and contributing author); and *Intro-*

ductory Botany Workbook. Dr. Uno was a program officer in the Division of Undergraduate Education at the National Science Foundation in 1998-2000 and serves on the editorial boards of four science and science education journals. He was awarded honorary membership by the National Association of Biology Teachers in 2001 and was its president in 1995. He became a fellow of the AAAS in 2000, and he has received one national, two state, and three university-level teaching awards. He has taught nearly 10,000 undergraduates and has led many faculty development workshops for university and secondary science instructors. He has served on the Board of Directors for the American Institute of Biological Sciences, and he recently published an article on botanical literacy in the *American Journal of Botany*. Currently, he is principal investigator for the NSF-funded "Introductory Biology Project," which focuses on the undergraduate freshman biology course; he is also co-PI on another NSF project dealing with professional development for high school teachers; and he is a member of several committees of the College Board that are revising advanced placement science courses.

STAFF

DIRECTOR

Jay B. Labov is senior advisor for education and communication for the National Academy of Sciences and National Research Council. He also has been the study director for the NRC reports *State Science and Technology Policy Advice: Issues, Opportunities, and Challenges* (2008); *Enhancing Professional Development for Teachers: Potential Uses of Information Technology* (2007); *Linking Mandatory Professional Development to High Quality Teaching and Learning: Proceedings and Transcripts* (2006); *Evaluating and Improving Undergraduate Teaching in Science, Mathematics, Engineering, and Technology* (2003); *Learning and Understanding: Improving Advanced Study of Mathematics and Science in U.S. High Schools* (2002); *Educating Teachers of Science, Mathematics, and Technology: New Practices for the New Millennium* (2000); *Transforming Undergraduate Education in Science, Mathematics, Engineering, and Technology* (1999); *Serving the Needs of Pre-College Science and Mathematics Education: Impact of a Digital National Library on Teacher Education and Practice* (1999); and *Developing a Digital National Library for Undergraduate Science, Mathematics, Engineering, and Technology Education* (1998). He has served as director of the Committee on Undergraduate Science Education, deputy director for the NRC's Center for Education, and oversees the National Academy of Science's efforts to improve the teaching of evolution in the public schools. Prior to assuming his position at the NRC Dr. Labov was a member of the biology faculty for 18 years at Colby College (ME), where he taught courses in introductory biology, topics in neurobiology, animal behavior, mammalian and human physiology, and tropical ecology. He was elected as a fellow in education of the American Association for the Advancement of Science in 2005. He received his bachelor's degree from the University of Miami and masters' and Ph.D. degrees from the University of Rhode Island.

Cynthia Wei is a National Academies Christine Mirzayan Science & Technology Policy Fellow, working at the National Academy of Sciences with Dr. Jay Labov (through December 2011). Dr. Wei recently completed an AAAS Science & Technology Policy Fellowship at the National Science Foundation in the Division of Undergraduate Education, where she worked on a wide range of issues in STEM education, focusing primarily on biology education and climate change education. She has diverse teaching experiences as a K-6 general science teacher, high school biology teacher, and university instructor, and she has taught college-level courses including introduction to zoology, introduction to animal behavior, field animal behavior, and biology and society. She is also an active member of the Animal Behavior Society's Education Committee, and co-organized a recent education workshop, Vision, Change, and the Case Study Approach. Dr. Wei received a dual-degree Ph.D. in zoology and ecology, evolutionary biology, and behavior from Michigan State University, and a B.A. in neurobiology and behavior from Cornell University. She also was a postdoctoral research associate at the University of Nebraska, Lincoln's Avian Cognition Laboratory.

C

Brief Biographies of Presenters and Panelists

Bruce Alberts, a prominent biochemist with a strong commitment to the improvement of science and mathematics education, serves as editor-in-chief of *Science* and as one of President Obama's first three Science Envoys. Dr. Alberts is also professor emeritus in the Department of Biochemistry and Biophysics at the University of California, San Francisco, to which he returned after serving two six-year terms as the president of the National Academy of Sciences (NAS).

During his tenure at the NAS, Dr. Alberts was instrumental in developing the landmark National Science Education standards that have been implemented in school systems nationwide. The type of "science as inquiry" teaching we need, says Dr. Alberts, emphasizes "logical, hands-on problem solving, and it insists on having evidence for claims that can be confirmed by others. It requires work in cooperative groups, where those with different types of talents can discover them—developing self confidence and an ability to communicate effectively with others."

Dr. Alberts is also noted as one of the original authors of *The Molecular Biology of the Cell*, a preeminent textbook in the field now in its fifth edition. For the period 2000 to 2009, he served as the co-chair of the InterAcademy Council, a new organization in Amsterdam governed by the presidents of 15 national academies of sciences and established to provide scientific advice to the world.

Committed in his international work to the promotion of the "creativity, openness and tolerance that are inherent to science," Dr. Alberts believes that scientists all around the world must now band together to help create more rational, scientifically based societies that find dogmatism intolerable.

Widely recognized for his work in the fields of biochemistry and molecular biology, Dr. Alberts has earned many honors and awards, including 16 honorary degrees. He currently serves on the advisory boards of more than 25 non-profit institutions, including the Gordon and Betty Moore Foundation.

Cynthia M. Beall (see Appendix B)

Paul Beardsley (see Appendix B)

Spencer Benson is the director of the Center for Teaching Excellence, associate professor in the Department of Cell Biology and Molecular Genetics and an affiliate associate professor in the Department of Curriculum and Instruction at the University of Maryland, College Park. Dr. Benson has served as a consultant for Project 2061, the Quality Undergraduate Education (QUE) initiative, the Coalition for Education in the Life Science (CELS), Science Education for New Civic Engagement and Responsibility (SENCER), and the Center for Advancement of Stem Education (CASE). He has been involved in numerous K-16 education initiatives at the University of Maryland including an on-line Master Program in the Life Sciences for high school biology teachers. He is past chair of the Undergraduate Education Committee of the American Society of Microbiology (ASM), past chair of ASM's Div-W (Teaching), and interim chair member of ASM's International Education Committee. He is a founding member of the International Society for the Scholarship of Teaching and Learning (ISSoTL) and the ASM sponsored Biological Scholars Program. In the 2002 he was named the CASE-Carnegie Maryland Professor of the Year award and in 2011 he was awarded the ASM Carski Teaching award. Dr. Benson has been an AP Biology exam reader (six years), test item reviewer, cochair of the AP Biology Redesign Commission (2006-2007), a member of the AP Biology Review Advisory Panel (2008), and cochair of the AP Biology Curriculum, Development and Assessment Committee (2008-2012).

V. Celeste Carter is a program director in the Division of Undergraduate Education (DUE) of the National Science Foundation (NSF). Dr. Carter received her Ph.D. in microbiology from the Pennsylvania State University School of Medicine in 1982 under the direction of Dr. Satvir S. Tevethia. She completed postdoctoral studies in the laboratory of Dr. G. Steven Martin at the University of California at Berkeley. She joined the Division of Biological and Health Sciences at Foothill College in 1994 to develop and head a Biotechnology Program. She served as a program director twice in the Division of Undergraduate Education as a rotator. Dr. Carter accepted a permanent program director position in DUE in 2009; she is the lead program director for the Advanced Technological Education (ATE) Program in DUE.

Betty Carvellas retired in 2007 after teaching science for 39 years at the middle and high school levels. She was a founding member of the National Academies Teacher Advisory Council (TAC) and currently serves as the Teacher Leader for the TAC. Her interests include interdisciplinary teaching, connecting "school" science to the real world, and bringing the practice of science into the classroom. Throughout her career, she traveled extensively on her own and with students. Her professional service includes work at the local, state, and national levels. She served as co-chair of the education committee and was a member of the executive board of the Council of Scientific Society Presidents and is a past president of the National Association of Biology Teachers. Included among her awards are the Outstanding Science Teacher-Vermont (1981), Presidential Award for Excellence in Mathematics and Science Teaching (1984), and a Christa McAuliffe fellowship. In 2001, she was selected for an NSF program, Teachers Experiencing Antarctica and the Arctic, and she has since participated in seven research expeditions in the Arctic. In 2008, she was designated a lifetime National Associate of the National Research Council of the National Academies. She received her B.A. from Colby College, her M.S. from the State University of New York at Oswego, and a Certificate of Advanced Study from the University of Vermont.

Amy L. Chang has served the American Society for Microbiology (ASM) Education Board since the 1980s. The ASM is one of the oldest and largest life science organizations, representing 38,000 members worldwide. About 60 percent of the members are microbiologists employed as faculty, staff, administrators, researchers, and students at colleges and universities. The Board advances the ASM's mission to educate individuals at all levels in the microbiological sciences. ASM is a voluntary organization. ASM members, serving as leaders and scientific experts, work in concert with a professional staff to sponsor programs, advance the ASM mission, and ensure stability.

Under her leadership, the Board is responsible for educators and faculty programs including the (i) annual Conference for Undergraduate Educators; (ii) professional development program in science teaching and science education research (Biology Scholars and Faculty Programs); and (iii) *Journal of Microbiology and Biology Education* and digital resources for microbiology. In September 2000, the Board was bestowed with the Presidential Award for Excellence in Mentoring Underrepresented Minorities in Science, Math, and Engineering Sciences.

The Board sponsors for students (i) national research fellowships; (ii) the Annual Biomedical Research Conference for Undergraduate Minority Students (ABRCMS); and (iii) professional development programs for graduate students and postdoctoral scientists in grantsmanship, publishing, presentations, teaching and mentoring, ethics, and career planning.

James P. Collins (see Appendix B)

William (Bill) Galey is director of graduate and medical education programs at Howard Hughes Medical Institute (HHMI). He directs HHMI's programs to enhance biomedical science graduate education and scientific training of medical students. He directs the HHMI Medical Research Fellows Program, which provides opportunities for medical students to engage in a year of intensive year of research. Dr. Galey was intimately involved in the HHMI partnership with the Association of American Medical Colleges known as Scientific Foundations for Future Physicians (SFFP), which sets out the scientific competencies needed by physicians to practice medicine in the 21st century. Graduate education efforts under Dr. Galey's direction include the Med into Grad Program, supporting efforts of graduate programs to graduate Ph.D.s with a strong understanding of medicine. Dr. Galey's group also administers HHMI's Gilliam Fellowship Program, supporting individuals committed to creating a more diverse professoriate. A new program known as the HHMI International Student Dissertation Research Fellowship Program has been initiated to support international graduate students during their dissertation research. Dr. Galey and his group also developed and conducted a highly successful partnership with the NIH to integrate graduate training in the physical and computational sciences with the biomedical sciences in a program known as Interfaces. Dr. Galey holds a Ph.D. from the University of Oregon Medical School, and was a fellow of Harvard University. After a brief period in the pharmaceutical industry, he joined the University of New Mexico School of Medicine (UNMSOM) where he conducted research and taught medical and graduate students. Dr. Galey was active in the development of problem-based learning and other educational innovations while a faculty member at the University of New Mexico. He also held various administrative positions including associate dean for graduate studies and interim dean for research at UNMSOM before joining HHMI in 2002.

David M. Hillis is the Alfred W. Roark Centennial Professor in Natural Sciences at the University of Texas, in the Section of Integrative Biology.

He uses a mix of molecular and computational approaches to study problems in molecular evolution and biodiversity. He is also actively involved in reform of science education at the university level. Dr. Hillis led the effort to reorganize the biological sciences at the University of Texas, and then served as the first director of the new School of Biological Sciences. He currently serves as the director of the Dean's Scholars Program for the College of Natural Sciences, as well as director of the Center for Computational Biology and Bioinformatics. He is a John D. and Catherine T. MacArthur Fellow, and has been elected to membership of the National Academy of Sciences and the American Academy of Arts and Sciences.

Kristin Jenkins (see Appendix B)

John Jungck is vice president of the International Union of Biological Sciences and editor of *Biology International*. He is the Mead Chair of the Sciences at Beloit College and professor of biology. Dr. Jungck has specialized in mathematical molecular evolution, image analysis, history and philosophy of biology, and science education reform. In 1986, he co-founded the BioQUEST Curriculum Consortium, a national consortium of college and university biology educators devoted to curricular reform across the nation. It promotes quantitative, open-ended problem solving, collaborative learning, peer review, research, and civic engagement/social responsibility. He is a Fulbright Scholar (Thailand), a Mina Shaughnessy Scholar, a fellow of the National Institute of Science Education, and a fellow of the American Association for the Advancement of Science. He teaches genetics, cellular and developmental biology, evolution and topics courses on bioinformatics, Darwin, and science and culture.

Susan Kassouf, program officer, has served in different capacities at the Christian A. Johnson Endeavor Foundation since 1999, first working with the Educational Leadership Program and now working more directly with grantees. After receiving her B.A. from Hampshire College and a Ph.D. in German Studies from Cornell University, she taught on the faculty at Vassar College.

Jay Labov (see Appendix B)

Joseph C. LaManna is the current president of the Federation of American Societies for Experimental Biology (FASEB). Dr. LaManna is also a professor of physiology and biophysics, neurology, neurosciences, and pathology at the Case Western Reserve University (CWRU) School of Medicine in Cleveland, Ohio. He is the former chair of the Department of Anatomy at CWRU (1993-2008). He received his undergraduate degree in biology at Georgetown University in Washington, DC, in 1971. He earned a Ph.D. in physiology and pharmacology from Duke University in Durham, NC, in 1975.

He has been involved in cerebrovascular research for more than 30 years. Research conducted in his laboratory is concerned with energy demand, energy metabolism, and blood flow in the brain. The role of these mechanisms in the tissue response to pathological insults such as stroke, cardiac arrest and resuscitation, and hypoxia is being actively investigated. His most recent research has centered on cerebral angiogenesis and the role of hypoxia-inducible factor-1 in physiological adaptation to hypoxia, neuroprotection, and ischemic preconditioning. He has authored or co-authored more than 200 research papers and review chapters.

Dr. LaManna currently serves on the editorial boards of the *Journal of Applied Physiology*, the *Journal of Cerebral Blood Flow and Metabolism*, and *Brain Research*. He is an active member of multiple scientific societies including the Society for Neuroscience (Program Committee, 2002-2005); American Physiological Society; International Society for Oxygen Transport to Tissues (Executive Committee, 1986-1989; 1995-1998; 2000-2003, President, 2009); AAAS; International Society of Cerebral Blood Flow and Metabolism (Board of Directors, 2007-2011, Secretary, 2011-2017); Association of Anatomy, Cell Biology and Neuroscience Chairs (Executive Board 2002-2006); and American Association of Anatomists (Public Affairs Committee Chair, 2002-2007).

He served as a regular member of the NIH Neurology B-1 Study Section, and is a current member of the Brain Injury and Neurovascular Pathologies (BINP) study section.

Joe Levine earned his Ph.D. from Harvard, has taught lecture and field courses at Boston College and Boston University, and currently teaches Inquiry in Rain Forests, a graduate field course for teachers through the Organization for Tropical Studies. His popular scientific writing has appeared in trade books, in magazines such as *Smithsonian* and *Natural History*, and on the web. Following a fellowship in Science Broadcast Journalism at WGBH-TV, he served as science correspondent for National Public Radio's Morning Edition and All Things Considered, and helped launch Discovery Channel's *Discover Magazine*. He served as scientific advisor to NOVA for programs including Judgment Day, and as science editor for the OMNI-MAX films Cocos: Island of Sharks and Coral Reef Adventure, and for several PBS series, including The Secret of Life and The Evolution Project. He has led seminars and professional development workshops for teachers across the United States, Mexico, Puerto Rico, the U.S. Virgin Islands, Indonesia, and Malaysia. With Kenneth Miller, he co-authors *Biology* (Pearson Education), the most widely-used high school biology program in the United States. This book is a frequent target of anti-evolution activity because of its thorough and curriculum-wide coverage of evolutionary biology. It was the flashpoint for the *Kitzmiller v. Dover Area School District* trial and the Cobb County, GA textbook sticker case. Dr. Levine currently serves on the Board of Overseers at the Marine Biological Laboratory in Woods Hole, and the Board of Visitors of the Organization for Tropical Studies.

Ross Nehm is an associate professor of science education and evolution, ecology and organismal biology at The Ohio State University. He received a Ph.D. in integrative biology at the University of California, Berkeley, an Ed.M. in science education at Columbia University, and a B.S. in geology (paleobiology) at the University of Wisconsin-Madison. Major honors include a CAREER award from the National Science Foundation, an outstanding instructor award from Berkeley, and a college-wide mentoring award from the City University of New York. In 2006 he was named an Education Fellow in the Life Sciences by the National Academy of Sciences. He publishes widely on topics relating to evolution, scientific thinking, student learning, and assessment methodologies.

Robert T. Pennock is professor of history and philosophy of science in Lyman Briggs College, and also holds appointments in the Philosophy Department, the Department of Computer Science, and the Ecology, Evolutionary Biology and Behavior Graduate Program. He is a member of and the Briggs faculty liaison to the Center for Ethics and Humanities in the Life Sciences. He is one of the Co-PIs of the BEACON Center for the Study of Evolution in Action, an NSF Science and Technology Center.

Dr. Pennock's research involves both experimental and philosophical questions that relate to evolutionary biology and cognitive science, such as the evolution of altruistic behavior, complexity, inference, and intelligence. He uses digital evolution (Avida as well as evolving neural networks) to investigate the emergence of intelligent behavior. Rather than trying to build intelligent systems from the top down, he is interested in investigating how such systems evolve from the bottom up. His Evolving Intelligence (EI) group has focused on the evolution of elements of intelligent behavior, including phenotypic plasticity, short-term and associative memory, environmental information processing, purposeful movement control, and cooperation. His research in these areas has been published in numerous journals and featured in *Discover*, *New Scientist*, *Science Daily*, *Slashdot*, *US News & World Report*, and many other national and international periodicals.

Dr. Pennock is also involved with various national initiatives to support undergraduate education about evolutionary biology and more generally about the nature of science. He leads the NSF-funded Avida-ED project, which develops and assesses software and curricular materials to use evolutionary computation to help teach these concepts. He was an expert witness in the *Kitzmiller v. Dover Area School Board* case that ruled that Intelligent Design creationism is not science, but sectarian religion, and that teaching it is the public schools is unconstitutional. He was the co-founder and first president of the citizens action group Michigan Citizens for Science.

Dr. Pennock also studies the relationship of epistemic and ethical values in science. Scientific methodology itself comes with tacit norms that govern appropriate professional behavior. His work in this area deals with what he calls the scientific virtues, which is a new way to think about some issues in responsible conduct of research. He is currently writing a book on this topic.

In recognition of his education and public outreach work, he was named a fellow of the American Association for the Advancement of Science, a Sigma Xi National Distinguished Lecturer, and a National Associate of the National Academies of Science, and has received a number of awards, including the National Center for Science Education's Friend of Darwin Award (2003) and the American Institute of Biological Sciences Outstanding Service Award (2009).

Jaclyn Reeves-Pepin is the executive director of the National Association of Biology Teachers (NABT). NABT has long been an advocate for maintaining scientific integrity in the classroom, and the teaching of evolution has been one of the association's main tenants for more than 70 years. As the executive director, Reeves-Pepin coordinates all programs at NABT to ensure alignment with NABT's mission to empower teachers and support students.

As this relates to evolution education, Reeves-Pepin schedules speakers and presentations at the NABT Professional Development Conference to ensure that evolution and the teaching of evolution are major themes at this event; she provides assistance producing the evolution-themed issue of the journal, *The American Biology Teacher*; she assists with the NABT Evolution Education Award (sponsored by AIBS and BSCS); she composes letters and statements made on behalf of the NABT Board of Directors, including statements regarding legislation and science textbook; she interacts with teachers to help them access local and national resources to assist in the teaching of biology; and she works with partner organizations to promote evolution-based resources and opportunities to both the NABT and larger biology educator communities.

Judy Scotchmoor is assistant director of the University of California's Museum of Paleontology (UCMP) in Berkeley, overseeing the museum's education and outreach efforts. Ms. Scotchmoor received her B.S. in biological sciences at UC Berkeley in 1966 and then proceeded on to a long teaching career, primarily at the middle school level. She began her career at UCMP as a volunteer in the fossil prep lab in 1993, before joining the staff the following year. Taking advantage of her K-12 experiences, she soon initiated teacher professional development workshops and curriculum development focusing on evolution, paleontology, the geosciences, and their intersection reflected in the biodiversity that we see today. Today she is the project coordinator of three award-winning websites: The Paleontology Portal, Understanding Evolution, and Understanding Science. She is the editor/author of three books to support K-16 teaching: *Evolution: Investigating the Evidence*, *Learning from the Fossil Record*, and *Dinosaurs: The Science Behind the Stories*. Ms. Scotchmoor was the recipient of the Joseph T. Gregory Award for outstanding service to the Society of Vertebrate Paleontology in 2004, was the recipient of the American Institute of Biological Sciences Education Award in 2006, was named an American Association for the Advancement of Science Fellow in 2009, and was elected as a Fellow of the California Academy of Sciences in 2011. She serves on the boards of AIBS and Impact100 Sonoma.

Mark D. Schwartz is associate professor of medicine at New York University (NYU) School of Medicine. After studying medicine at Cornell University and training in internal medicine at NYU, Dr. Schwartz was awarded a Bowen-Brooks Fellowship by the New York Academy of Medicine to study medical education innovation in Israel and Holland, and then completed a General Internal Medicine Fellowship at Duke University. At NYU he was selected as a Robert Wood Johnson Generalist Physician Faculty Scholar. He has been a primary care physician in urban underserved settings for 20 years.

Dr. Schwartz has studied primary care workforce issues since the 1980s and recently completed a national study of influences on student interest in internal medicine. His health services research focuses on how primary care workplace characteristics impact physician stress and burnout and, subsequently, quality of care and medical errors. He also leads a Veterans Administration study of how educational interventions for health professionals improve patient outcomes.

Since 1995, Dr. Schwartz has led NYU's General Internal Medicine Fellowship Program and established its Master's of Science in Medical Education program. He directs NYU's NIH Clinical Research Training Program and leads its Master's of Science in Clinical Investigation Program. He also directs the Fellowship in Medicine and Public Health

Research. NYU recently named him director of translational research education and careers in its Clinical Translational Science Institute. The Association of Clinical Research Training awarded him its Distinguished Research Educator award in 2008. In his practice, educational leadership, research, and scholarship, Dr. Schwartz has focused on the need to improve health and health care of vulnerable, urban poor populations.

Maxine Singer attended the New York City public schools and graduated from Swarthmore College (A.B., 1952, with high honors) and Yale University (Ph.D., biochemistry, 1957). She joined the National Institutes of Health as a postdoctoral fellow in 1956 and received a research staff appointment two years later. She was chief, Laboratory of Biochemistry, National Cancer Institute, 1980-1987, where she led 15 research groups engaged in various biochemical investigations. She became president of the Carnegie Institution in 1988 and President Emeritus in 2002. She retains her association with the National Cancer Institute as Scientist Emeritus. At Carnegie she established (in 1989) the Carnegie Academy for Science Education (CASE) whose goal is to enhance learning of science and math for DC public school teachers and students. Now she works actively in several CASE projects. Dr. Singer's research contributions ranged over several areas of nucleic acid biochemistry and molecular biology, including chromatin structure, the structure and evolution of defective viruses, and enzymes that work on DNA and its complementary molecule, RNA. Around 1960 she collaborated intensely with her NIH colleague Marshall Nirenberg in the elucidation of the genetic code. In recent years, her foremost contributions have been in studies of a large family of repeated DNA sequences called LINEs that are "jumping genes" and are interspersed many times in human DNA. Researchers elsewhere found that LINE-1 insertions into, for example, a gene whose product is required for blood clotting are associated with cases of hemophilia. She has published more than 130 scientific papers and several books on molecular genetics (with Paul Berg). Throughout her career, Dr. Singer has taken leading roles influencing and refining the nation's science policy, often in realms having social, moral, or ethical implications. In 1975 she was one of the organizers of the Asilomar Conference on Recombinant DNA. Among countless other roles in service to science and humankind, she was chairman of the editorial board of *Proceedings of the National Academy of Sciences*, 1985-1988 and the chair of the Academies' Committee on Science, Engineering and Public Policy (COSEPUP) (2000-2005). She was a member, Board of Directors, Johnson & Johnson (1990-2002), and the Yale (University) Corporation, 1975-1990. She was elected to the National Academy of Sciences in 1979 and to membership in the Pontifical Academy of Sciences in 1986. In 1992 she received the National Medal of

Science, the nation's highest scientific honor bestowed by the President of the United States, "for her outstanding scientific accomplishments and her deep concern for the societal responsibility of the scientist." She was awarded the National Academy's Public Welfare Medal in 2007.

Dr. Singer served on the panel that wrote the first *Science and Creationism* document for NAS (1983-1984). She also served on the panel that did the NAS' report entitled *Teaching about Evolution and the Nature of Science* (1998).

Paul Strode teaches international baccalaureate (IB) biology at Fairview High School in Boulder, Colorado. Dr. Strode has a Ph.D. in ecology and environmental science (2004) from the University of Illinois at Urbana-Champaign and holds a science education master's (1996) from the University of Washington (Seattle). After completing a B.S. degree in biology, chemistry, and secondary education (1991) from Manchester College (IN), Dr. Strode taught biology and chemistry at Hazen High School in Renton, Washington. Dr. Strode grew up in the small college town of North Manchester, Indiana, where he spent a lot of his free time on his bike and playing with friends next to and sometimes in the Eel River. His natural love of the biological sciences was fully realized in his high school freshman year in his biology class. His teacher, Harvey Underwood, was trained as a forest ecologist and had his students spend a lot of time outside collecting and identifying insects and leaves. Dr. Strode has no memory of learning evolutionary theory in high school or college, even though his college zoology professor had published several papers on the evolution/creationism dichotomy. Dr. Strode also has no memory of learning how science works until he learned-by-doing in his doctorate program (after teaching high school science for eight years). He has published peer-reviewed scientific articles, middle school science textbook chapters, and a book titled *Why Evolution Works (and Creationism Fails)* with Matt Young. Dr. Strode was interviewed about bird migration and climate change on NPR's "All Things Considered" (May 3, 2006) and about teaching evolution on KGNU Denver/Boulder's "How on Earth" (June 28, 2011).

Gordon E. Uno (see Appendix B)

Marlene Zuk received her undergraduate degree at the University of California, Santa Barbara, and her Ph.D. at the University of Michigan. After doing postdoctoral work at the University of New Mexico, she joined the faculty in biology at the University of California, Riverside, where she is now a professor. Her research interests include behavioral ecology, sexual selection, and the evolution of host-parasite interactions. Recently Dr. Zuk has become interested in how behavior can influence the rate of evolution.

Most of her work has used insects, although she also has studied birds. She is interested in communicating science to the public and has written three books for general audiences: *Sexual Selections: What We Can and Can't Learn About Sex from Animals*, published in 2002; *Riddled with Life: Friendly Worms, Ladybug Sex, and the Parasites that Make Us Who We Are*, published in 2007; and *Sex on Six Legs: Lessons on Life, Love, and Language from the Insect World*, which was released in 2011.